FINE ARTS, MUSIC AND LITERATURE

# ADVANCES IN ARTS BIOMECHANICS

# FINE ARTS, MUSIC AND LITERATURE

Additional books and e-books in this series can be found
on Nova's website under the Series tab.

FINE ARTS, MUSIC AND LITERATURE

# ADVANCES IN ARTS BIOMECHANICS

GONGBING SHAN
AND
PETER VISENTIN
EDITORS

nova
science publishers
New York

**Library of Congress Cataloging-in-Publication Data**

ISBN: 978-1-53615-641-6

*Published by Nova Science Publishers, Inc. † New York*

# CONTENTS

# PREFACE

Attitudes toward the role of the Arts in society are as varied as they are numerous, ranging from 'the Arts are a (nice) diversion' to 'while many things may be necessary to make living possible, it is the Arts that make life worth living'. In the last few decades, research in the areas of psychology, neuroscience and sociology have underpinned the positive health benefits of the arts as well as artists' roles in encouraging innovative thinking and the development of novel perspectives for the betterment of society. But, artists work within the existence of boundary conditions. For even the most subversive of artistic creations, the act of subversion is inherently referential to the condition that is being subverted. Furthermore, adaptation of "our modes of perception borrowed from the sensations" are hindered by the reality that "every word in the language refers to our ordinary perceptions" (Neils Bohrs, *Nature Supplement*, April 14, 1928), and is thus mediated by both the design capacities and limitations of the human form and the ways in which we express that experience. Hence, by examining multiple facets of the commonality of bodily experience Arts Biomechanics explores the human capacity to translate between the perspectives of self and other.

Arts Biomechanics as a field of study uses the tools and methods of science to understanding the bodily experience within artistic creation. Desired aesthetic outcomes are influenced by: 1) how an artist might use metaphor or symbolism rooted in bodily experience, 2) how an artist might experience the act of creation, 3) how a performance of an artistic act may be better understood through analysis of physical skills necessary for that act, 4) how the act of challenging creative boundaries might challenge the capacity of the body, and 5) how representations and perception of the human condition are mediated by technology – "the medium is the message" (Marshal McCluhan, *Understanding Media: the Extensions of Man*, 1964). In all of these areas Arts Biomechanics allows a deeper discourse, rooted in the commonality of bodily experience, exploring the relationships between modes of perception and com

*Advances in Arts Biomechanics* has twelve chapters covering music, dance, and the visual & media arts. It is organized in three parts. The first part, music research, has articles covering fundamental theory, proof of principles, musicians' health, motor control/learning and its application in practice. Many of these are meaningful for both pedagogy and performance. The second part, dance articles, examines the biomechanics of dancing, kinematics and kinetics, injury prevention, and biomechanical foundations of intercultural representations of gender roles. Finally, two visual & media arts articles discuss motion capture use in performance and artistic creation as well as its communication to an audience, linking bodily gesture to the performative act whether on stage or on canvas. These innovative articles represent advances in thinking regarding biomechanics and the arts. By their very nature, using the tools and methods of science to better understand the visual and performance arts, all are interdisciplinary. We hope that the included articles challenge and inspire researchers and artists in the pursuit of transdisciplinary ways of knowing and creating in the arts.

# SECTION 1. MUSIC

In: Advances in Arts Biomechanics
Editors: Gongbing Shan and Peter Visentin
ISBN: 978-1-53615-641-6


*Chapter 1*

# THE ROLE AND MANAGEMENT OF TENSION IN PEDAGOGICAL APPROACHES TO PIANO TECHNIQUE

***M. A. Michèle Wheatley-Brown[1,*], Gilles Comeau[1], PhD and Donald Russell[2], PhD***

[1]Pedagogy Research Laboratory, School of Music,
Faculty of Arts, University of Ottawa, Canada
[2]Department of Mechanical and Aerospace Engineering
Faculty of Engineering and Design, Carleton University, Canada

## ABSTRACT

Mastering the piano to play with freedom and without physical limitations has long been a topic of great interest and importance to pianists. This is borne out in the numerous pedagogical approaches to piano technique. Playing without tension is considered an important key to technical freedom among many of these approaches. However, conflicting advice on the role and management of tension abounds. The purpose of this study is to investigate the meaning and management of tension in current pedagogical approaches.

To do this, the language used to define and describe the positive and negative role of tension was collected from five pedagogues who have developed approaches that reflect current trends in piano technique: Barbara Lister-Sink; Dorothy Taubman; Thomas Mark; Fred Karpoff; and Alan Fraser.

The data was then analyzed by examining both the meaning and management of tension in each approach. This study showed that the authors view tension very differently—either as an impediment to motion, and thus a quality that must be banished—or as an essential component of control and activity, and thus a quality that must be carefully harnessed and managed.

Three reasons for this confusion were identified: there is an inaccurate understanding of the anatomical and biomechanical principles of the role of tension in piano technique;

* Corresponding Author's E-mail: wbrowns@telus.net

the authors waver between using scientific, common, and invented terms to describe tension; and, there are challenges describing the fine degree of muscular control needed to play the piano freely. By recognizing where the problems in the varied meanings and management of tension exist, this study represents an important first step for the pedagogical community to bring more clarity to how we teach the role and management of tension in piano technique.

## INTRODUCTION

Since the invention of the early pianoforte over three hundred years ago to the modern day piano, pedagogues and pianists have developed approaches aimed at mastering piano technique.

The literature on this subject is vast and varied: the finger action school of Muzio Clementi (1803), Carl Czerny (1844), and Charles Hanon (1873); the relaxation and arm weight school of Rudolph Breithaupt (1905) and Tobias Matthay (1903, 1932); the scientific-based investigation of Otto Ortmann (1925, 1929); the empirical approaches developed by Abby Whiteside (1929, 1955), Seymour Bernstein (1981, 1991), Seymour Fink (1992, 1994); and more recently, the numerous approaches designed to address playing-related injuries from pedagogues such as Dorothy Taubman (Taubman, Golandsky, Schnitzer, Urvater, Bloomfield, & Yaguspsky, 2001), and Barbara Lister-Sink (1996, 2008).

Pedagogues have pursued piano technique from different perspectives over the piano's lifetime. Since the beginning of the twentieth century, many piano approaches are based on the conviction that technical mastery provides the key to injury-free technique, physical and musical freedom—and for many, the guiding principle is that this is achieved by avoiding excessive muscular tension.

### Detrimental Tension

Throughout much of the literature on piano technique, tension figures as a major culprit in limiting agility and causing injury. Louis Adam (1758-1848), a pioneer of the Paris Conservatoire tradition, wrote early method books for piano that served as one of the first pedagogical resources of this venerable institution (Parakalis, 2001). In a letter advising young piano forte players, (1829), Adam wrote about the constrictions that come from tension: "…for all muscular tension blocks freedom of movement and makes it impossible for the hands to extend and contract the fingers properly, which one needs to be able to do at all times" (as cited by Rosenblum, 1998, p. 195).

Almost two hundred years later, Thomas Mark, author of *What Every Pianist Needs to Know About Their Body* (2003), an approach designed to teach musicians how to move in accordance with an accurate understanding of the anatomical design of the body, conveys the same prevailing view held by many pedagogues and pianists in the following statement: "Piano playing that is accomplished by high-quality movement, in which each part contributes its proper share with no tension, will be free, expressive, and secure" (p. 5).

This sentiment is also summed up in the title of Barbara Lister-Sink's popular approach to injury-free technique, *Freeing the Caged Bird* (1996, 2008), a metaphor for the freedom that comes from releasing the physical and mental entrapment of tension. The belief that tension is a major contributing factor to playing-related injuries and technical limitations even inspired one of Britain's leading piano pedagogues, Carola Grindea (1914-2009), to found the *International Society for the Study of Tension in Performance* (*http://www.isstip.org/*). This society provides symposiums, clinics, and articles to raise awareness of how tension impacts physical problems in performance and seeks solutions to managing tension for a healthy and free technique.

## Disaccording Voices

Not all pedagogues have embraced this notion that tension must be avoided. Fixation of joints with a degree of tension was an important aspect of Rudolph Breithaupt's approach to piano technique. Indeed, he believed that "a gifted individual will, of his own accord and unconsciously, exert a certain muscular tension as required by the part or situation" (Breithaupt, p.99, 1909). Seymour Bernstein (1981), popular pedagogue, writer, and clinician criticizes the idea that pianists must eschew all tension from their playing. He states that it is important to recognize the different attributes of tension by categorizing all tension as either: 'organized tension', 'controlled tension', 'muscular tension', and 'emotional tension'. Similarly, Alan Fraser, author of *The Craft of Piano Playing* (2003, 2006), and *Honing the Pianistic Self-Image* (2010) contests the negative view of tension: "We must do away with this fear that tension will interfere with, or limit motion. Without tension there is no life at all" (Fraser, 2003, p. 38). These views differ dramatically from the more prevalent belief that tension is a major impediment to achieving technical freedom. Clearly, these conflicting views represent a significant source of confusion for pianists and pedagogues learning how best to approach the physical aspect of piano technique.

## Relaxing as an Antidote to Tension

When pedagogues do agree that harmful tension must be avoided, they often advise relaxation as a remedy. Tobias Matthay, one of the most influential pedagogues of the early 20th century (Gerig, 2007), wrote extensively of the importance of 'relaxation'. He sums up his thoughts on relaxation in his landmark work, *The Act of Touch in All Its Diversity* (1903): "Really complete relaxation may possibly be physically unattainable, but the term thoroughly conveys what has *practically* to be striven for. For it is the completest possible *Relaxation* of all muscles that should be passive that forms the real 'secret' of all good tone-production (including Agility) at the Pianoforte" (p. 22). Matthay's belief that 'relaxation' was the key to the technical mastery required for good tone production shaped subsequent pedagogical approaches and is a concept that is firmly entrenched in much of the literature today (Gustafson, 2007). Indeed, of the many ways to avoid, reduce, or eliminate unhealthy tension, the ability to relax immediately after the finger makes contact with the key is presented as an important element of a healthy and skilled technique in modern day approaches (Lister-Sink, 1996, 2008; Riley, Coons, & Marcarian, 2005).

Despite his belief that the term 'relaxation' described an optimal muscular state, Matthay later regretted the use of this term as pianists had misunderstood this to mean throwing their arms on to the piano with absolutely no control (Bernstein, 1981). Other pedagogues such as

Abby Whiteside, Arnold Schultz, and Seymour Fink have also shared concerns that the term 'relaxation' has led pianists to develop dangerous ideas of how to move to play the piano. Indeed, Dorothy Taubman (2001) has raised concerns that encouraging relaxation actually contributes to harmful tension. Alan Fraser (2010) suggests that pianists have become so accustomed to 'over-relaxation' that "we simply don't perceive its detrimental effect" (p. 232). He believes that encouraging relaxation can put the skeletal structure out of "alignment" and muscles cannot initiate movement efficiently. These opinions contrast starkly with the view that relaxation is an important antidote to harmful tension and add to the confusion the pianist faces when considering how best to address harmful tension.

## Science-Based Pedagogical Approaches

In an effort to legitimatize and offer a more logical explanation of the mechanics of piano technique, some pedagogues have turned to science as a basis for their approaches. Taubman believed that it was a "lack of knowledge" that stymied piano virtuosity—not talent, and therefore it was essential "that we avail ourselves to scientific procedures" (Taubman, et al., 2001, chap 1). Lister-Sink defines her technique from a "neuromuscular and biomechanical viewpoint" (1996, 2008). Fred Karpoff has developed an award-winning approach to technique that comes from an understanding of "how to merge healthy biomechanics with dynamic music making" (2009, p. 6). Despite these claims of using science as a foundation for piano technique, underlying contradictions of the role and management of tension in piano technique continue to surface. For example, Taubman and Mark discourage situations causing opposing muscles to contract simultaneously around a joint. Yet, from a biomechanical standpoint, this is a necessity in any movement (Ackland, Elliott, & Bloomfield, 2009). Lister-Sink's method also presents misleading information in suggesting that muscles only serve to fixate joints —not initiate movement (Fraser, 2010). Recently, scholars of both piano pedagogy (Wristen, 2000) and mechanical engineering (Vant, 2007) conducting research into the biomechanics of piano technique have also raised concerns that there is a widespread misunderstanding of muscle and joint function in the piano pedagogical literature. Even though pedagogues have embraced a more scientific perspective in analyzing and observing motions to play the piano, Wristen, piano pedagogue and researcher at the University of Nebraska-Lincoln states that "many of these pedagogues based their recommendations on an incomplete of even incorrect knowledge of anatomy and human motion" (p.55), ironically, creating situations that produce rather than prevent injury. In a study entitled *Driving point impedance measurements during piano playing,* Vant (2007) also points to the incongruity between the two fields of biomechanics and piano pedagogy: "Expertise and language in the two fields are considerably different and thus the two fields are divided by a great expanse…" (p. 34). Despite the stated inclusion of scientific principles in pedagogical approaches to piano technique, these are not always presented or articulated accurately, and the recurring confusion continues.

## Research Problem

Although pedagogues today share the same goals of playing the piano freely, without physical limitation, and without injury—there is much conflicting advice with regards to muscular effort and joint function—particularly surrounding the concept of tension. Even when pedagogues claim to base their approaches on scientific principles, we are no closer to agreement on how tension figures in piano technique. With no consensus on the meaning, the role, and the ways of dealing with tension, there is little constructive debate and discussion, and the pedagogy of piano technique cannot advance. This is an acute problem. There is a need to better understand how contradictions and inconsistencies surrounding the meaning and management of tension may contribute to the confusion in the pedagogy of piano technique. Therefore, this study poses the following questions:

1. What are the consistencies and inconsistencies in the meaning and management of tension among pedagogical approaches?
2. Does the pedagogical definition of tension align with the scientific definitions?

In answering these questions, this study brings to light the issues of misunderstandings in language surrounding the confusion in piano technique. This is an important first step in bringing more clarity and consistency to how language is used to convey the concept of tension in piano technique. Ultimately, a more precise, consistent, and common manner of communicating tension will lead to a more productive discussion and debate of its role and management in piano playing.

# METHODOLOGY

To answer the above questions, this study conducted a content analysis of the usage of 'tension' in current pedagogical approaches. The following five approaches were selected for this study due to their focus on the physical aspect of piano technique: 1. *The Taubman Technique*, [DVD] (Taubman et al., 2001)[1]; 2. *Freeing the Caged Bird* with Barbara Lister-Sink [DVD] (Wingsound, 1996, 2008); 3. *What Every Pianist Needs to Know about the Body* [DVD and book] (Mark, 2003)[2]; 4. *3-D Piano with Fred Karpoff,* [DVD and workbook] (Karpoff, 2009); 5. *The Craft of Piano Playing* [DVD, book, and workbook] (Fraser, 2003) and *Honing the Pianist Self-Image* ([book] Fraser, 2010). Collecting the data entailed a detailed reading and viewing of the books and DVDs. Each time 'tension' appeared in the material, it was systematically recorded into tables designed to assemble a detailed examination of its use. The tables include the following categories: Context (to show how the term is employed: i.e., 'tension' used when describing curled fingers); Definition/Description (to show how the author defines/describes the term); Terms (to provide words associated with

[1] Taubman's assistant at the time of the production of the DVD, Edna Golandsky, demonstrates Taubman's principles and reads from Taubman's lecture notes throughout the DVD.

[2] Body Mapping evolved from William Conable's discovery in the mid-1970s that music students who mapped accurate anatomical information of their bodies were able to adopt principles of the Alexander technique more quickly. Barbara Conable developed Body Mapping into a course that trains musicians to move more freely and expressively by learning an accurate internal representation of the structure, size, and function of the body's anatomy.

the term; i.e., 'stiff' and 'fixed' associated with tension); Role (to determine if the concept is presented by the authors as contributing positively or negatively to piano technique); Summary (to provide a synopsis of the use of the term). This data was then amalgamated into a second set of tables that summarized underlying causes, effects, and ways to address both the positive and negative roles of tension in piano performance.

There were two components to the data analysis: 1) consistency of pedagogical usage, and 2) scientific accuracy. First, the consistency of the pedagogical usage of the concepts was examined by tallying the frequency of references to the positive and negative roles. These numbers indicated whether the authors consistently or inconsistently associate the concept with a positive or negative role. The terms the authors use to refer to tension in both its negative and positive role were then organized into "scientific", "common", and, "invented" categories.[3]

These categories emerged during the data collection. At times, the authors explain and describe how tension relates to muscle, joint function and movement with scientific terms, thus the inclusion of 'scientific terms' in the data collection. At times, the authors use these terms in a colloquial sense with broad meaning, thus the inclusion of 'common terms' in the data collection.

Lastly, at times, the authors blended the scientific and common terms to come up with their own interpretation of the meaning of tension and relaxation, thus the inclusion of 'invented terms' in the data collection.

Allocating each of the terms to scientific, common, and invented categories provided an overview in how the authors used the concept. To complete the analysis of pedagogical usage, the summary tables were consulted to determine how each author addressed the positive and negative aspects of tension.

The second component of the data analysis examined the scientific accuracy of the pedagogical usage of tension. Biomechanical resources were consulted to determine a standard definition of tension.

The scientific definition provided a benchmark to compare the pedagogical usage of tension and allowed for the identification of specific areas of accuracies and inaccuracies in how tension is presented in the pedagogical approaches.

## RESULTS

Tension figures as a central concept in each of the author's approaches. Indeed, a cursory reading would perhaps not have revealed how prevalent tension appears in all of the approaches. From the systematic and numerous records taken for the initial data collection tables it was possible to produce a chart, (Table 1.1), showing tension as having only a negligible positive impact on piano technique (14/157) with much more attention to its negative aspects (143/157).

[3] By "invented" we mean a term created or re-defined by an author for use within the context of a particular work. By "scientific", we mean a term that is clearly defined, accepted and used with consistently in science and engineering. By "common" we mean a generally vague or ambiguous term that is used in casual conversations or in a situation where sufficient background does not exist to use the term precisely.

**Table 1.1. Frequency of references to tension as playing either a positive or negative role**

| Author | Positive | Negative |
|---|---|---|
| Lister-Sink | 2/22 | 20/22 |
| Taubman | 0/10 | 10/10 |
| Mark | 0/69 | 69/69 |
| Karpoff | 0/12 | 12/12 |
| Fraser | 12/44 | 32/44 |
| Total references from all approaches | 14/157 | 143/157 |

## Positive Role of Tension

In its positive manifestation, two of the authors agree that tension provides necessary joint stability. The difference lies in how this tension is managed. The language the authors use to describe positive aspects of tension in Table 1.2 sheds some light on these differences. Lister-Sink's method revolves around developing kinesthetic awareness of 'minimal' muscular tension needed to stabilize joints at the instant the finger makes contact with the key and learning to relax and release all of this tension immediately after it has performed this necessary task. Although Fraser also draws attention to the importance of cultivating sensory awareness to muscular tension, he devotes more attention to its positive role in all aspects of piano playing. He argues that in addition to providing important stability to joints, tension also serves to control, to move, and to provide important finger activity. Rather than completely eliminate tension, Fraser's approach focuses on how to best harness the positive aspect of the opposing forces created by tension. Recognizing the negative associations with the term 'tension', Fraser proposes using the term 'muscle tonus' to describe tension's positive attributes, defining this as the "appropriate muscular effort needed to hold bones in place" (2010, p. 241).

**Table 1.2. Scientific, common, invented terms to describe tension as playing a positive role**

| Author | Scientific terms | Common terms | Invented terms[4] |
|---|---|---|---|
| Lister-Sink | Muscle contraction;<br>Stabilize; | *Quantifiers:* Minimal;<br>"Just the right amount of tension—no more, no less" (chap. 2).<br>*Qualifier:* Stability | N/A |
| Taubman | N/A | N/A | N/A |
| Mark | N/A | N/A | N/A |
| Karpoff | N/A | N/A | N/A |
| Fraser | Co-contraction;<br>Muscle tonus;<br>Tensile strength; | *Quantifiers:* Appropriate; Necessary; Healthy; Reduced; Minimal:<br>*Qualifiers:* Muscular effort;<br>Natural muscle tonus<br>Stability | Internal tension;<br>Transformative tension; |

[4] These are a blending of personal, scientific, and common terms.

The relationship between stability and muscular tension is less clear among the other authors. Taubman and her assistant Golandsky describe the joints as serving as "fulcrums", but Golandsky states that "to fixate or mold is wrong, however much the amount…even a little bit would be wrong" (Taubman, et al., 2001, chap. 5). Similarly, Mark discourages any degree of fixation. He believes that movement must be organized around a stable bony structure, stating that this is a "stability achieved without tension" (p. 112). Despite speaking of this necessary 'stability', Mark also suggests that using the term 'stable' can create an erroneous idea that movement must initiate around a fixed place and thus lead to dangerous tension. Karpoff is less categorical in dismissing muscular activity for stability and advises students to "respect the antagonistic relationship" (2009, Unit 1, chap. 17) to provide 'important stability'. Once again, the pianist confronts conflicting advice. From explanations of joints acting as fulcrums, avoiding any type of fixation, to a vague degree of muscular activity providing necessary stability, the pedagogues present a muddled picture of how muscles and joints work to play the piano.

## Negative Role of Tension

The terms all of the authors use to describe the negative role of tension underscore its detrimental role in hampering healthy and free movement. As shown in Table 1.3, terms such as 'lock' 'clenching' 'cramped' and 'paralysis' convey tension as resisting movement. In almost all cases, the authors attribute playing-related pain and injury to excessive tension. The amounts of tension that can bring about these negative consequences range from 'any' to 'excessive'. Clearly, in the view of these pedagogues, tension figures as a major impediment to healthy and free technique. Given this view, the pedagogues, to varying degrees, focus attention on reducing unnecessary tension from piano playing.

## Differing Ways of Reducing Tension

Instructions to 'release' tension are found in all of the approaches. Although united in the view that releasing harmful tension leads to more fluid and healthy piano playing, there is inconsistency when describing how to execute this release. Learning to release tension by consciously controlling the activation and complete release and relaxation of muscular contractions is at the core of Lister-Sink's approach. She likens the neuromuscular system to a light switchboard—when signals to the brain turn off, muscles relax. It is this state the pianist must return to, immediately and constantly, after the muscles contract to stabilize joints when making contact with the key. Instructing students to "relax the entire arm muscles completely, a state of release to which the pianist must always return" (Lister-Sink, 1996, 2008, chap. 1), Lister-Sink conveys full relaxation and release as vital to avoiding a build-up of tension. Conversely, Golandsky states that we cannot command muscles to relax and activate and even suggests, paradoxically, that relaxation is another guise of tension: "if people are very relaxed, they have a tremendous amount of tension" (2001, chap 10). She believes tension can only release by cultivating a quality of movement without undue tension. Mark takes this idea further by encouraging movement based on an accurate internal representation of the body's anatomical design.

Broad instructions to 'release tension' accompany his descriptions of anatomy relevant to playing the piano. Although Fraser also draws attention to the important relationship of anatomy and movement, he advocates a completely different approach to addressing tension. Believing that a structurally secure hand and fingers will allow undue tension to release in the rest of the body, Fraser promotes a hand structure that is organized around strong and pronounced three bony arches of the forefinger, thumb, and metacarpophalangeal (MCP) joints. In contrast, the other authors advocate a very loose, quiet, and natural hand position to avoid tension.

In fact, Taubman believes the very hand position of a pronounced MCP joint that Fraser recommends as a remedy for tension is actually a cause of tension. To add to the confusion, Karpoff describes 'relaxed motion', and a 'very relaxed hand' as desirable—a stark contrast from Golandsky's belief that relaxation is another form of tension and Fraser's 'activated' hand and fingers.

**Table 1.3. Scientific, common, invented terms to describe tension as playing a negative role**

| Author | Scientific terms | Common terms | Invented terms |
|---|---|---|---|
| Lister-Sink | Muscular contraction | *Quantifiers:* Chronic; Excess; Accumulative; Any; Unnecessary; *Qualifiers:* Resistance; Paralysis; Locks; Dangerous; Harmful; | Inner tension; Hidden tension; |
| Taubman | Static muscular activity | *Quantifiers:* Too much; Slightest; Tremendous; *Qualifiers:* Muscular effort; Clenching; Fixating; Tightening; | Dual muscular pull |
| Mark | Static muscular activity | *Quantifiers:* Excessive; Chronic; Continual; Constant; Repeated Small amount; Compensatory *Qualifiers:* Muscular effort; Short, narrow muscles; Underlying; Upward; Stiff; Strain; Gripping; Fixed; Holding; Squeezed Stabilize; Counteract; | Dual muscular pull |
| Karpoff | Muscular tension | *Quantifiers:* Accumulative; Extra; Excess; Reduce; *Qualifiers :*Cramped; Held; Lock; Unbalanced; | N/A |
| Fraser | Co-contraction Excessive muscle tonus | *Quantifiers:* Extra Excess; Negative; Unwanted Undesired; Undue; Useless Increased; Inappropriate; Overly; Any; Too much; Persistent; Superfluous; Counterproductive Compensatory; *Qualifiers:* Muscular effort; Effortful; Opposing forces; Constraint; Destabilize; Stabilize; Immobility; Blocking; Inhibit; | Internal Inner |

We can see that the pedagogues view the role of tension differently, resulting in contradictory recommendations. Either seeking its total absence to facilitate ease and freedom, or discerning just the right amount to stabilize joints and empower technique, these varied ways of dealing with tension can be traced to how the authors define tension. In Lister-Sink's case, when tension is defined as a necessary but temporary muscle contraction in providing stability in joints, the approach focuses on a minimal contraction that is then completely released. If tension is viewed as too much muscular effort that inhibits movement, the authors focus on a piano technique that ranges from reducing tension to completely eliminating tension. In contrast, Fraser's concession that there are two forms of tension—healthy and unhealthy—results in an approach that revolves around creating structural support in the hand and fingers with muscular tension. Fraser's attempt to differentiate between healthy and unhealthy tension by using either the terms 'muscle tone' or 'tension,' though perhaps important steps in providing more clarity in how tension figures in piano technique, is not consistently used within Fraser's own approach; nor is it a term that is used by the other authors. Nonetheless, Fraser's proposal to use different terms to refer to healthy and unhealthy tension recognizes that there is a problem in how the term 'tension' is used. Indeed, the differing interpretations found among the authors leads to conflicting advice in piano technique. To come to a clearer understanding of the meaning of tension, we now turn to the scientific definition and consider the accuracy of the pedagogical usage of this concept.

## Analyzing Accuracy

In mechanics, the word *tension* refers to a *force* (a vector, that is, it has both magnitude and direction) that tends to pull on the object on which the force acts[5] (Brickmann, Frobin, Leivseth, 2002). The application of a force to an object does not imply that the object will be in motion as other forces applied to the object may create a situation in which the net force applied to the object is zero.

When the object is a muscle several other basic principles of mechanics and biomechanics are also important. Skeletal muscles generally act to cause two bones to rotate relative to each other (a joint). Motion will only occur in a joint if the net effect of all of the muscles and any external forces applied to the bones is unbalanced. Muscle is connected to bone through tendons. Tendons (and muscles) in the body cannot support compression - only *tension*.[6] As a direct result of this, almost all joints have more than one muscles acting across the joint - the contraction of one muscle will tend to cause the joint to rotate in one direction while contraction of the other muscle will tend to cause the joint to rotate in the opposite direction. In the situation where all of the forces tending to rotate the joint are appropriately balanced the joint will not move and neither muscle will change length. A muscle that is contracting but not changing length is said to undergo an *isometric* contraction. In the situation where all of the forces applied to a joint result in motion of the joint one (or more) muscles will lengthen and one (or more) muscles will shorten. An active muscle that is *shortening* is generating a force *(tension)* that acts in a direction that tends to increase the movement of the joint. The contraction in such a muscle is described as *concentric*. A muscle acting on the other side of the joint that is contracting will be generating a tensile force that

[5] The opposite of tension is compression - a pushing force.

[6] Obeying the often stated axiom in introductory mechanics: "You can't push on a rope."

tends to reduce the movement of the joint. The contraction in such a muscle is described as *eccentric*.[7] (See Vant, 2007).

The force applied to (or by) a muscle is not solely dependent on whether or not the muscle is being contracted. A muscle that is not contracting can still be under tension; for example, when external forces arising from other muscles or outside of the body act to lengthen the muscle, the *passive* material properties of the muscle result in the muscle stretching and, as a result, in a spring-like force that resists the applied tension.

There is only one length, the so-called, *rest length*, of the muscle where, without a contraction, it generates no force. As a result, in almost all positions, every muscle in a limb is generating some *passive tension* as a result of being stretched but not as a result of it contracting (See McMahon, 1984).

A muscle contraction (the active generation of force) may be voluntary or involuntary. A voluntary contraction is one that results from intent - a decision to move. An involuntary contraction of a skeletal muscle, is generally the result of a reflex. Whether a contraction is voluntary or involuntary does not change the biomechanics of the muscular contraction. In both cases, the contractile elements within the muscle work to create *tension* in the muscle. *Tension* arising from the contraction of a muscle is often described as being *active* (as opposed to passive).

The use of the term tension is properly limited to describing the force generated by the muscle and not the cause of any contraction that might be involved or the resulting motion of the muscle or any joints across which it may act.

In the act of playing the piano muscles in the arm will be generating both active and passive tension. At any particular time, muscles in the arm and hand will be contracting concentrically and others contracting eccentrically; postural muscles, for example, in the back and shoulders may be contracting but not changing length - isometric contraction[8]; most will be contracting voluntarily but involuntary activity will be contributing to the development of *tension* in the muscles. The level of voluntary contraction in a muscles will be the result of many factors (including sensory and emotional factors) and as a result may be more or less than is required, from a mechanical point of view, to achieve the task of performing a piano work.

Four points from the scientific use of *tension* provide a framework to examine and analyze the accuracy of the pedagogical usage of tension: (1) all muscular contractions—concentric, eccentric, and isometric—produce *tension,* (2) the degree of *tension* varies according to the amount and duration, (3) a degree of *tension* is always present, albeit a very small amount and (4) *tension* describes the force generated by a muscle and does not describe a joint.

The following examines how the pedagogical usage of tension aligns with these scientific points.

---

[7] Note that in addition to isometric, concentric and eccentric, muscles are often described as having an *isotonic* contraction. The word isotonic describes a situation in which the tension generated by the muscle is constant. The adjectives isometric, concentric and eccentric are mutually exclusive but unrelated to whether or not a contraction is isotonic.

[8] In some situations, it may be the case that, in holding a chord, a muscle maintains a contraction without changing length (an isometric contraction) but it is important to realize that just because the finger tips are not being moved does not mean that the muscles associated with the movement of the finger tips are not changing length. It is a simple task to hold down a piano key and move all of the joints in the arm.

## 1. All Muscular Contractions—Concentric, Eccentric and Isometric—Produce Tension

All of the pedagogues agree that muscular contractions produce tension, however, with the exception of Fraser, they exclusively associate tension with an isometric contraction that serves to maintain a stationary position and impede the motion of a joint. Thus, the majority of terms referring to tension found in the pedagogical approaches do not consider tension's role in producing and controlling movement. Terms such as fixate, tighten, hold, lock, and stiffen, repeatedly emphasize a static quality of tension. Fraser stands out as the only author who allows that tension is present in all types of muscular contractions. He argues that the supportive function that comes from the simultaneous contraction of muscles can free movement, describing tension as "two opposing muscle groups working purposefully one against the other, to achieve musical ends" (2010, p. 38). To avoid confusion for his readers, and knowing that tension is viewed so negatively, Fraser proposes reserving the word 'tension' to describe a contraction that serves to inhibit movement, and 'muscle tonus' for necessary muscular activity. Although it is true that tension can resist movement, the language found in the pedagogical approaches narrows the description of tension to a product of isometric contraction and fails to acknowledge its role in producing and controlling movement from concentric and eccentric muscle contraction.

## 2. The Degree of Tension Varies According to the Amount and Duration of Muscular Contraction

There is a tendency among the pedagogues to present extremes of either too much or a total absence of tension with little possibility of a changing amount and duration of tension to accommodate changing situations. The pedagogues share the common concern that poor movement patterns and poor skeletal alignment will lead to an increase of magnitude and duration of tension. Extra effort is required to move; therefore, tension accumulates and is generally regarded as too much. According to Mark, tension decreases when using just the right amount of muscular effort to execute the task at hand. In this manner, there is an acknowledgement that tension increases and decreases with effort. However, there are challenges in articulating an optimal degree of *tension* that will provide stability but not impede movement. In Lister-Sink's words: "It is a question of finding just the right amount of tension—no more, no less" (1996, 2008, chap. 2). Karpoff's instruction to 'respect the antagonistic relationship' implies that there is a fine but vague muscular activity the pianist should strive for. Yet, when Golandsky states "the slightest tension will limit motion" (2001, chap. 5), the pianist is left with the impression that all tension must be eliminated.

## 3. Tension Is Always Present in Healthy Muscles, Albeit a Very Small Amount

The language found in the pedagogical approaches vacillates between acknowledgements that tension is always present or absent. Fraser states: "Without tension, there is no life" (2003, p. 38), pointing to the fact there is always a degree of muscular activity that gives the skeletal structure support. Even though his overarching approach is that of finding an optimal balance of tension, some of his advice implies that any degree of tension is detrimental. In describing the transmission of arm weight, Fraser warns, "If any tension remains in the elbows, it will interfere with the free transmission of energy from your arm to the key" (2010, p. 207). As shown in the analysis of consistency, instructions and comments from the other pedagogues are also often directed towards playing with a total absence of tension. While qualifying tension with terms such as minimal and excessive suggest the constant presence of *tension,* the pedagogues' directions to release 'all' tension mislead and misinform. In conveying *tension* as something that can be totally eliminated, the pedagogical use of tension does not align with the scientific fact that a degree of *tension*, albeit at times very small, is always present in healthy muscles.

## 4. Tension Describes the Force Generated by a Muscle and Does Not Describe a Joint

Much confusion arises from the use of the word tension to describe what is happening at a joint. When a single muscle is contracting it will generate *tension* that will tend to cause (or resist) the rotation of one or more joints.

If a second contracting muscle also generates *tension* that acts to prevent this motion (Taubman's "dual-muscular pull" or, from a biomechanical point of view, "co-contraction"), the joint may remain stationary. Pedagogues often describe this situation as a joint under tension, but this only serves to confuse the meaning of the word. Clear use scientific terminology, would use the term "stiffness" (or its opposite "relaxation") to describe joints which are subject to multiple opposing muscular forces. The roll that co-contraction and joint stiffness plays in piano playing is complicated but it is clear that in many situations undue amounts of co-contraction are not beneficial.

This analysis of accuracy shows that although there are steps to move towards a more scientific use of tension, there are still inaccuracies in the pedagogical usage. The pedagogues predominantly use tension in the context of isometric contractions that most often results in unwanted and harmful resistance to movement. Some of the authors suggest that no tension is needed at all to stabilize joints, leading to a use of terms that qualify tension as an all or nothing state.

Even Fraser, who shines a light on the positive effects that come from muscular tension, and advocates for an approach to playing that uses tension 'intelligently', also recommends relegating the term to its negative manifestations of inhibiting motion. Although this effort to distinguish between the positive and negative consequences of tension is an admirable attempt at bringing more clarity to this issue, Fraser does not consistently use 'tension' to refer to solely its negative attributes within his own work.

This highlights the confusion in narrowing the definition of tension to only impeding movement. A more accurate description of tension would consider its role in producing and controlling movement and encompass a variety of physical movements needed to play the piano results with differing degrees of *tension*. Instead, the pianist is left with the overriding impression that tension serves only to impede movement at joints and is a quality that must be banished, not managed.

## CONCLUSION

The extensive collection of data on the language used to describe the role and management of tension from the selected authors revealed significant problems with inconsistency and inaccuracy. An analysis of the language found various reasons for these problems: there is an incomplete or inconsistent understanding and explanation of anatomical and biomechanical principles; the authors freely interchange "scientific", "common", and "invented" terminology; and the authors struggle to articulate the opposing dualities of stability and flexibility that comes from managing tension. Note that the terms "stability" and "flexibility" also exhibit problems similar to those identified in this paper for "tension".

The language presents a confusing and inaccurate picture of the role of tension in joint and muscle function. As shown, much of the confusion surrounds how tension figures in joint stability. When viewed as necessary to providing stability, the management of tension ranges from its complete elimination to finding the correct amount. When viewed as unnecessary to providing stability, the pedagogues focus on playing without any degree of tension. The belief that tension does not provide stability or that stable joints do not play a role in movement clearly contradicts biomechanical principles. It is possible that this emphasis on banishing tension from piano playing comes from concerns that most pianists move with too much tension (or perhaps more accurately joint stiffness) that impedes free movement. In addition, there is some confusion in the subjective and objective nature of tension. Lister-Sink makes the point that even though a minimal amount of tension is needed, it will not be felt thus. Rather than giving directives that reflect the true mechanics of tension in movement, advice reflects the common belief that all tension is unnecessary and the subjective idea of "harmful" tension.

This tendency to address tension with such broad strokes is illustrated in the use of colloquial and invented terms. Nonetheless, as shown, there are times when an attempt is made to explain tension in scientific terms. Frequently switching between the terms, the concepts take on different meanings depending on how they are being used. For example, as a common term, tension can mean a chronic state of tight muscles that inhibits movement ("stiffness") and becomes a primary cause of playing-related-injury; as an invented term, it can mean something that must be 'emptied'; and lastly, as a scientific term, tension may refer to a muscle contraction that is a necessity in all intentional movement. This extraordinary array of scientific, common, and invented use of the word tension underscores the complexity of describing the mechanics and the perception of physical movement. However, it is this wide range of terminology that often contributes to ambiguous, vague, conflicting and confusing advice.

In part, the reasons for this inaccurate and inconsistent representation of tension come from the difficulty in describing "just the right amount of tension", and "just the right amount of relaxation", to provide "just the right amount of joint stability". Neither too tense, nor too relaxed, it is this state of stability and flexibility that often proves difficult to explain. Lacking a terminology that adequately conveys these contrasting qualities, the language tends to focus on the desirable qualities of a relaxed muscle, flexible joints, and the absence of tension, but often overlooks the necessary qualities of stiffness and stability associated with tension. Of note, Fraser has begun to flesh out this discussion of the complex and subtle degrees of relaxation and tension with his in-depth descriptions of muscular and joint function. The next step will be to engage the pedagogical community at large in an exploration of a language that accurately describes the dual opposing effects that come from the delicate balance of tension and relaxation. An open and frank dialogue about the challenges of describing this state will, first, acknowledge the true nature of tension; and secondly, address the issue of the management of tension, rather than focusing on its complete elimination. Exploring a vocabulary that communicates this balance will bring both more consistency and accuracy in the understanding and management of tension.

As this research evolves, there will be pedagogical implications. Pedagogy books may adopt anatomical terms and use more discretion in the use of terms such as tension and relaxation. Exploring terms that describe the appropriate balance of tension and relaxation and of movement at the joints will lead to a more informed and open discussion of piano technique. The language used by pedagogues to discuss the mechanics of technique amongst themselves will not necessarily be the same language used to communicate concepts to young students; but it is important that language aimed at a younger audience is rooted in accuracy with a common understanding of its meaning. This scope of this study was limited to pedagogical material directed at more advanced pianists. Although this allowed for a more intense investigation, research into similar issues surrounding the use of language in novice pedagogical material would be a valuable contribution. Although this study focused on the concept of tension, other core concepts such as arm weight, hand shape, joint stiffness and finger shape also surfaced as areas of contradictory advice within the pedagogical literature. These concepts may be the subject of further investigation in forthcoming papers.

This study has identified inconsistencies and inaccuracies in the language used to describe piano technique and sheds light on some of the issues contributing to these problems. To address these problems and issues, changes are needed. If piano technique is promoted as being based on anatomical and biomechanical principles, then the pedagogical community has an obligation to ensure that the language they use accurately reflects this. This will require a solid understanding of anatomy, joint function and muscle function along with a commitment to ensure that the language is consistent and accurate. This will require an effort to use language that clearly differentiates between the subjective experience and the reality, between metaphor and science, and this will require an open discussion of the challenges pedagogues face when describing the often opposing concepts that are the nature of piano technique. These represent some first steps in addressing the issues of how language is used. Ultimately, it is not just a matter of employing accurate language, but of developing a better understanding of the relationship of anatomy and mechanics of movement to piano playing. These are complex issues that require a consistent and accurate language. Moving towards a language that is clearly understood will facilitate advances in understanding the mechanics of technique. Piano technique need no longer remain a mystery, but a skill that can be explained logically and clearly to all students.

## References

Ackland, T., Elliott, B., & Bloomfield, J. (2009). *Applied anatomy and biomechanics in sport.* (2nd ed.). Australia: Authors.

Adam, L. (1829). *A Letter to a young piano forte player.* London: Hailes.

Bernstein, S. (1981). *With your own hands. Self-discovery through music.* New York, NY: Schirmer, Inc.

Bernstein, S. (1991). *20 Lessons in keyboard choreography. The basics of physical movements at the piano.* Milwaukee, WI: Hal Leonard.

Brickmann, P., Frobin,W., Leivsteh, G. (2002). *Musculoskeletal Biomechanics.* Stuttgart, Germany: Georg Thieme Verlag.

Breithaupt, R. (1905). *Die natürliche Klaviertechnik [The natural piano technique].* Leipzig, C.F. Kahnt, Nachfolger.

Breithaupt, R. (1909) *Natural Piano-Technic.* Translated by John Bernhoff. Leipzig: C.F. Kahnt Nachfolger.

Clementi, M. (1803) *Introduction to the Art of Playing the Pianoforte.* London: Clementi, Banger, Hyde, Colland, and Davis. Reprint in 1973, N.Y.: Da Capo Press.

Czerny, C. (1844). *The art of finger dexterity, op. 740.* Vienna: Mechetti. (Reprint in 1986, Sherman Oaks, CA.: Alfred Pub.

Fink, S. (1992). *Mastering piano technique: A guide for students, teachers, and performers.* Portland, Oregon: Amadeus Press.

Fink, S. (1994). *Mastering piano technique: A guide for students, teachers, and performers.*[DVD]. Portland, Oregon: Amadeus Press.

Fraser, A. (2003). *The craft of piano playing: A new approach to piano technique.* Lanham, MY: Scarecrow Press, Inc.

Fraser, A. (2006). *The craft of piano playing: A new approach to piano technique.* [DVD].Lanham, MY: Scarecrow Press, Inc.

Fraser, A. (2010). *Honing the pianistic self-image,* Novi Sad: Maple Grove Music Productions.

Gerig, R. (2007). (2nd edition). *Famous pianists & their technique.* Bloomington, IN: Indiana University Press.

Gustafson, A. E. (2007). *Tone production on the piano: The research of Otto Rudolph Ortmann* (Doctoral dissertation). Available from Proquest Dissertations and Theses database. (AAT 3274786).

Hanon, C. (1873, 1968). *The Virtuoso Pianist in sixty exercises.* New York, NY: Schimer. (Original work published in 1873).

Karpoff, F. (2009). *3-D piano with Fred Karpoff.* [DVD] Fred Karpoff (Producer). Available from http://3-dpiano.com/

Karpoff, F. (2009). *3-D piano with Fred Karpoff, Study Guide.* 3-D piano.

Lister-Sink (1996, 2008). *Freeing the caged bird. Developing well-coordinated injury-preventive piano technique with Barbara Lister-Sink* [DVD].Wingsound International. Available from www.freeingthecagedbird.com

Mark, T. (2003). *What every pianist needs to know about the body.* Chicago, IL: GIA Publications Inc.

Mark, T. (2003). *What every pianist needs to know about the body.* [DVD]. Chicago, IL: GIA Publications Inc.

Matthay, T. (1903). *The act of touch in all its diversity; an analysis and synthesis of pianoforte tone-production.* London: Longmans, Green and Co.

Matthay, T. (1932). *The visible and invisible in pianoforte technique.* London: Oxford University Press.

McMahon, T. A. (1984). *Muscles, Reflexes and Locomotion.* Princeton University Press.

Ortmann, O. (1925). *The physical basis of piano touch and tone.* London: K Paul, Trench, Trubner.

Ortmann, O. (1929). *The physiological mechanics of piano technique.* New York: Da Capo Press.

Parakilas, J. (2001). *Piano roles. A New History of the Piano.* New Haven & London: Yale University Press.

Riley, K., Coons, E., Marcarian D., (2005) *The Use of Multimodal Feedback in Retraining of Complex Technical Skills of Piano Performance,* Medical Problems of Performing Artists (6), 82-88.

Rosenblum, S. (1998). *Performance Practices in Classic Piano Music.* Bloomington and Indianaplis: Indiana University Press.

Stark, S. (1997). *The Stark Reality of Stretching.* Richmond, Canada. The Stark Reality Corp.

Taubman, D., Golandsky, E., Schnitzer, S., Urvater, E., Bloomfield, J., Yaguspsky, A. (2001). *The Taubman techniques.* Medusa, NY: Taubman Institute.

Vant, C. (2007). *Driving point impedance measurements during piano playing.* (Master's thesis). Available from Proquest Dissertations and Theses database. AAT MR33672, Carleton University. Canada.

Whiteside, A. (1929). *The pianist's mechanism.* New York, NY: G. Schirmer, Inc.

Whiteside, A. (1955). *Indispensables of piano playing.* New York, NY: Coleman Ross Company Inc.

Wristen, B. (2000). Avoiding piano-related injury: a proposed theoretical procedure for biomechanical analysis of piano technique. *Medical Problems of Performing, 15, 55-64.*

Wingsound International. (Producer). (1996, 2008). *Freeing the caged bird. Developing well-coordinated injury-preventive piano technique with Barbara Lister-Sink* [DVD]. Available from www.freeingthecagedbird.com

In: Advances in Arts Biomechanics
Editors: Gongbing Shan and Peter Visentin
ISBN: 978-1-53615-641-6


*Chapter 2*

# STUDY OF TWO STROKE TECHNIQUES WITH TRANSVERSE MOVEMENT OF THE RIGHT HAND ON CLASSICAL GUITAR USING SURFACE EMG

***Natalia Diaz Fernández de Monge*[1,*]**
***and Juan Granda Vera*[2]**
[1]Professional Classical Guitarist
[2]Human Movement Analysis Research Group
University of Grenada, Spain

## ABSTRACT

The purpose of this study is to examine expert playing patterns involving the index and middle fingers of the right hand of two professional classical guitar players using surface electromyographic (EMG) to analyse two different stroke techniques, rest stroke and free stroke, during the performance of a musical composition.

In the area of instrumental technique analysis, there are few studies that use EMG technology. The current study provides an empirical basis to help understand underlying motor control processes in classical guitar performance technique. Further it establishes a process for exploration of instrument technique during music performance.

Two case studies were conducted with professional guitarists who performed the classical guitar piece Estudio No. 5 by Francisco Tárrega (1993). Execution of the piece was recorded using EMG equipment and two video cameras. Subsequent analysis of the fingers involved the synchronisation of EMG data with each of the finger movements made during the performance of technically challenging sections of the music (e.g., transverse movement of the right hand or string crossing).

Results indicated a high degree of stability and consistency in terms of muscular activation, which demonstrated the excellent motor control of the expert guitarists. It is important to note that, each performer showed consistent patterns of motor control for each of the two repetitions of the musical passage even though a comparison of the two performers reveals different surface EMG patterns. These results indicate a measure of technique individualization during the performance of the technically challenging section

* Corresponding Author's E-mail: ndfdm@hotmail.com.

(transverse movement or string crossing) for the fingers (index and middle) and strokes (rest stroke and free stroke).

**Keywords:** music performance, instrumental technique, guitar, surface EMG, rest stroke, free stroke

## INTRODUCTION

Technical skills are required to perform any musical piece correctly. Expert-level technique serves to free the interpreter from physical constraints in order to allow for greater freedom of musical expression. Thus, technique should be considered a creative element instead of an end unto itself. It is a means of gaining confidence and control, and, through this, magnifying musical effects and improving performance outcomes (Deschaussées, 2002).

Typically, different master teachers have varying approaches to teaching guitar technique, resulting in considerable motor control differences, especially for right hand stroke techniques. Russell (1993) states that although the left hand movements of most advanced guitarists are generally standardised and some diversity exists in terms of different fingerings which may be used to reduce difficulty or for different musical preferences, there are considerable differences in the way the right hand is used. Each performer tends to adopt a personalized posture and technique. However, when discussing technique it is common to refer to the type of positioning of the hands, while what is truly and fundamentally important lies in coordinating multiple seemingly insignificant movements.

Consequentially, based on this, successful execution on the guitar can be accomplished in many ways and strings can be plucked in a variety of different manners. Decision-making with regard to the kind of stroke to employ is a primary methodological resource in teaching the instrument. Further, the manner in which a stroke is executed provides the means to develop proper technique and to facilitate the interpretation of musical compositions from different eras and styles. The two methods of plucking the string are: 1) the rest stroke, which involves resting the finger on the neighbouring string after a finger plucks a string, and 2) the free stroke, in which the finger does not rest anywhere, the finger only flexes (De Contreras, 1998; Duncan, 1980; Pujol, 1956; Schneider, 1985).

In general, musical technique and interpretation are founded on teaching methods governed by tradition and pedagogy which are commonly determined by the teacher's own artistic preferences (Visentin and Shan, 2011). In accordance with this, it is noteworthy that current pedagogical methods, which substantively focus on learning technique, are normally based on a series of practical tips which are either generally accepted because they are based on common sense or they have been proposed by teachers of recognised prestige. This reality and the small number of existing empirical studies on guitar technique underscore the relevance of the current study. The results of the study contribute objective information about how to play the guitar and motor control characteristics during performance.

In the field of guitar, one study comparing plucking techniques which did not employ EMG was conducted by Marques, Rosset-Llobet, Fonseca Marques, Gurgel and Augusto (2003). The study indicated rest stroke technique to primarily involve flexion and extension of the proximal phalanges of the three middle fingers. This distinguishes it from the free stroke technique, which involved incomplete flexion of the middle phalanges of the same digits. One of the conclusions reached by the authors was that rest stroke technique requires greater tension of the flexor muscles, because, after plucking a string, the finger then rests on the neighbouring string.

An additional study of classical guitar plucking techniques (rest stroke and free stroke) conducted by Granda, Barbero and Diaz (2006) examined the degree of muscle involvement using EMG recording for both expert and novice players. The results of the study showed differences in EMG activity between the index and middle fingers of the right hand and significant differences in muscle voltage levels between novices and experts.

In continuation to the work above, Granda, Barbero and Diaz (2007) conducted a study with guitar students at the Professional Music Conservatory of Melilla which determined levels of muscle voltage using surface EMG recording of the right hand. The rest stroke method was confirmed to involve lower levels of muscle voltage, allowing for better development at the beginner and intermediate guitar technique levels, both in terms of skill acquisition and retention.

EMG has also enabled researchers to examine specific techniques used with musical instruments other than the guitar. For example, Bejjani, Ferrara, Xu, Tomaino, Pavlidis, Wu, and Dommerholt (1989) published a research paper which compared three different techniques for playing the piano, confirming the utility of EMG as a means of distinguishing muscle activity for each of the techniques. The results of a study by Moore (1992) on the execution of trills on the piano showed that finger acceleration and EMG activity are presented as phases characteristic of the trill cycle. Furthermore, magnitude and coordination are dependent on speed of the execution of the trill, the dynamic level, and place where the finger hits the key. Another study with professional pianists to complement teaching techniques for piano was conducted by Montes, Bedmar and Sol Martin (1993). The study examined changes in EMG activity by measuring the short abductor muscle of the thumb. The results were used for continued study of the various execution techniques for this instrument. In the same vein, a study examining alternatives to help improve the quality of piano teaching was conducted by Granda, Barbero and Rodriguez (2004). Their results indicated that implementing a program based on information limitation appears to provide significant improvement in the motor control of hand actions, which results in a decreased voltage levels in the thumb extensor muscle and increased values for the little finger extensor, supported by kinaesthetic and proprioceptive information processes.

In terms of bowed string instrument technique, Moore, Hary and Naill (1988) analysed trills executed on the cello to determine the characteristics and speed of finger movement as these relate to the EMG activity of the muscles that control them, providing information for potential biofeedback studies on problems related to instrument practice. Although Fjellman-Wiklund, Grip, Karlsson and Sundelin (2004) used EMG as a measuring instrument to study disorders related to playing stringed instruments, their study showed that EMG can be used to analyse individual differences in the performance of a musical piece and to evaluate different methods of action. And lastly, the aim of the study conducted by Granda, Barbero and Rodriguez (2010) was to analyse and describe the kinematic patterns of the right arm and muscle voltage levels of the anterior deltoid muscles, finger flexors and extensor pollicis of a professional and highly proficient cellist during the interpretation of different sections of a piece of music. Their study demonstrated a stable movement pattern during the performance of a musical piece, and also showed variation in the kinematic values and muscle activity of the arm segments studied, which were offset by adjustments in other segments of the arm.

Another important finding of the study was the fact that the driving force for the right arm in a cellist's technique was the shoulder. The hand and fingers act to correct and absorb shock from the general movements stemming from the shoulder.

Using the abovementioned research as a foundation, the current research aims to determine muscle voltage levels through surface EMG recording during the performance of a musical piece by two highly proficient professional guitarists for two different stroke techniques in order to describe and analyse motor control patterns used in guitar performance which are the result of attaining expertise in the execution of motor skills for both plucking methods.

## METHOD

This study took the form of a case study experimental design with a single post-test. Each participant was considered a unique case, since the aim was to gain insight into the participants' performance behaviours (the patterns of activity of each participant) and to investigate possible intra-subject differences.

In the first phase of the experiment, an expert with over 40 years of experience as a teacher and more than 25 years dedicated to musical interpretation was asked what classical guitar piece would be most suitable for the study.

By this process, Estudio No. 5 by Francisco Tárrega (1993) was selected for the test protocol. The participants were then asked if they agreed that the piece would be useful for studying both right hand plucking techniques (rest stroke and free stroke) using the index and middle fingers.

This was necessary because, for many compositions, there can be a variety left and right hand fingering and plucking patterns. The participants agreed as to the appropriateness of the selected composition.

During this phase, two guitar virtuosos (experienced professionals) were asked to practice the aforementioned Estudio using both stroke techniques (rest stroke and free stroke) for 16 minutes a day (8 minutes of rest stroke practice, and 8 minutes of free-stroke practice) over the period of one month. They were also asked to take extreme care to keep their fingernails in optimum condition during their practice so that they would be in excellent condition on the day when the data was to be recorded.

| | Repetition 1 | Repetition 2 | Repetition 3 | Repetition 4 |
|---|---|---|---|---|
| Interpreter 1 | Free stroke | Rest stroke | Free stroke | Rest stroke |
| Interpreter 2 | Rest stroke | Free stroke | Rest stroke | Free stroke |

Figure 1. Order in which each of the guitarists interpreted the piece.

In the second phase, data were collected on the levels of muscle activity and finger actions. Prior to starting the experiment, all participants performed warm-up exercises, basically consisting of scales and passages from the Estudio. They were told to take as much time as they needed to warm up, to ensure they were fully prepared to perform the test piece.

In the data recording session, each guitarist played the Estudio 4 times: 2 times with the rest stroke technique and 2 times with the free stroke technique, starting with the technique each preferred (see Figure 1).

The guitarists interpreted the piece from memory in each of the four repetitions, and in each repetition the guitarists employed the same fingerings for the left and right hands. All of the soloists' performances were recorded using EMG equipment and two digital cameras. Surface EMG measurements were taken using the I-330 12-channel recording system (4 EMG), (J & J Engineering). Bipolar EMG recordings of extrinsic finger flexors were obtained using pairs of Ag-AgCl surface electrodes. EMG signals from the flexors of each finger were recorded on an ASUS F3J model laptop. Two Casio EXF1 digital cameras recording at 100 frames per second were used to record video and audio of the interpretation. These were positioned to facilitate subsequent analysis of the fingers involved in the study, the index and middle fingers, and for synchronisation of EMG data with each one of the moments (actions) of the fingers during the performance.

For surface EMG recording, electrodes were placed in accordance with the recommendations of Basmajian and Blumenstein (1989). Two pairs of electrodes were placed parallelly and longitudinally from the proximal to the distal insertion, so that each pair recorded tension generated by the actions of each of the fingers studied, with each compartment separated by 13 mm, the distance recommended by Shan and Visentin (2005) to avoid the cross-talk effect between the segments of muscle studied. This placement of surface electrodes did not allow the activity of the FDS finger superficial flexor to be separated from the FDP finger deep flexor. However, as force was applied in the distal phalanges of the finger position studied the deep flexor FDP was the main contributor to finger forces (Li, Zatsiorsky and Latash, 2000). The reference ground electrode was placed on the dorsal side of the forearm. All the EMG signals were sampled at 100 hertz, amplified, and then filtered using a high-pass filter at 10 hertz and a low-pass filter at 500 hertz, effectively resulting in band pass data of 10-500 hertz.

Each pair of electrodes recorded voltage generated by the muscles needed to produce the actions of each of the fingers studied. This allowed the researchers to verify the relationship between finger placement and muscle activity in order to analyse the motor control patterns of the finger segments. To ensure the validity and reliability of data recording, the activity of each finger was sampled separately with EMG recording software. The level of activation of performing fingers was verified, while the fingers which remained inactive showed minimum and unchanged tension.

For data analysis, a section of the Estudio was selected (Figure 2). In this passage, performers only use the index and middle fingers of the right hand. This section presented a series of relevant guitar technique challenges to be used in subsequent study of the action of the fingers which are the primary object of this study. The results for the technically challenging section selected correspond to transverse displacement of the fingers of the right hand, i.e., string crossing.

For the analysis of surface EMG measurements, the data collected was first synchronised with the videos recorded of the participants' performance using EMG recording software. Post-synchronisation, the videos were viewed to identify precisely which situations corresponded to the musical passages that were the focus of the study. Using this process, sections of EMG data that matched these video segments were then identified.

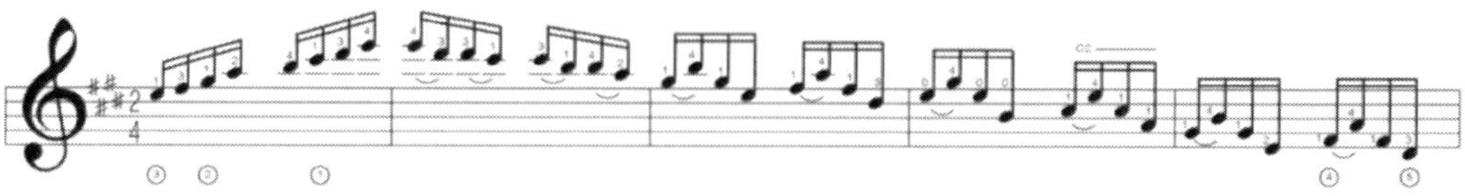

Figure 2. Selected passage (measures 3 to 7) of Estudio No. 5 by Francisco Tárrega (1993).

After extracting the audio-video recordings and the EMG for the identified musical passages, each section was viewed frame by frame in order to precisely and rigorously determine the EMG value for each moment of finger control during the execution of the stroke, providing the information that would later be used for analysis and interpretation of emergent behaviours in each situation analysed.

## RESULTS AND DISCUSSION

Surface EMG patterns for the index and middle fingers and the technique employed (rest stroke or free stroke) in each case are indicated below for some of the musical segments analysed. In each instance, the guitarists were performing a transverse hand movement, i.e., moving the fingers from one string to another laterally across the fret board. Transverse right hand movement requires correct left hand finger posture and therefore presents a technical challenge which is practiced from the initial stages of studying the instrument (Carlevaro, 1979).

### Participant 1

As illustrated in Figure 3, during the execution of identified technical passage using the rest stroke technique, the guitarist (participant 1) changed from the first string plucked with the middle finger to the second which was plucked with the index finger. When the second string is plucked, a minimal rise in voltage is observed which decreases as the index finger action ends and comes to rest on the neighbouring string. The higher voltage levels attained in the action of the plucking finger is the result of the intensity with which the guitarist executes this moment of the interpretation, in accordance with the requirements of the piece. In this same vein, in a study of pianists, Montes et al. (1993) noted that variation in EMG activity was related to stroke intensity.

In Figure 4, while crossing strings using the free stroke technique, the explanation for the surge in muscular voltage levels at the beginning of the action when approaching the string, and the reduction of these in the final moments of the action, may lie in the fact that when the free stroke is executed the guitarist does not have a point of support or reference to complete the action, and must apply greater control over the movement and speed of execution prior to plucking the string. To do otherwise an uncontrolled movement may occur in the impulse of the final action, and, as a consequence, an unintended musical effect may result.

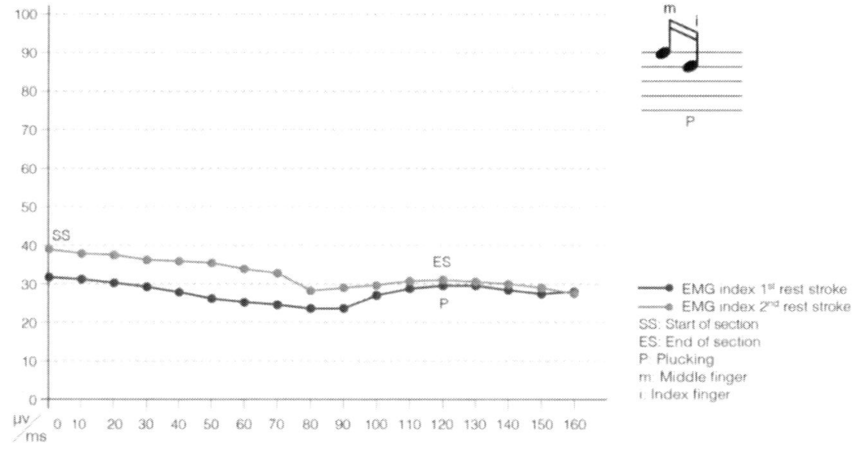


Figure 3. Surface EMG pattern for the index finger of the right hand, rest stroke, measure 5, second part.

This data corroborates the findings of Carlevaro (1979) on the mechanics of attack, which indicates that work is not completed with the attack itself, but must be complemented by counterforce as strong as or stronger than the original action to avoid abandoning the finger once plucking is completed. This is due to the fact that, in this musical situation, results indicate that the force applied is greater in the phase of impulse towards the string and, at the time of completion of the impulse, this may require compensatory action of the extensor muscles of the fingers to balance the action. This counteraction demonstrates the system may dictate different activation thresholds for each of the muscle groups in order to achieve greater control over the action (Feldman, 1998).

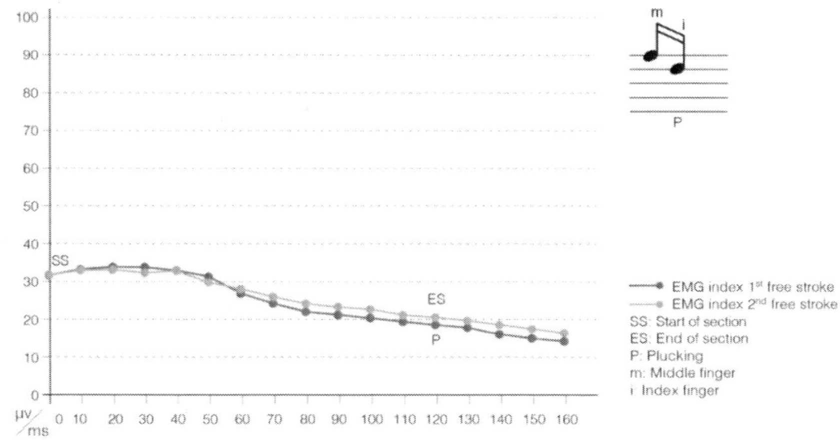


Figure 4. Surface EMG pattern for the index finger of the right hand, free stroke, measure 5, second part.

In Figure 5, during the execution of the technically challenging section using the rest stroke technique, the second string was plucked with the index finger, then the first string was plucked with the middle finger. At the moment of plucking, the lowest muscle voltage value was recorded in both repetitions.

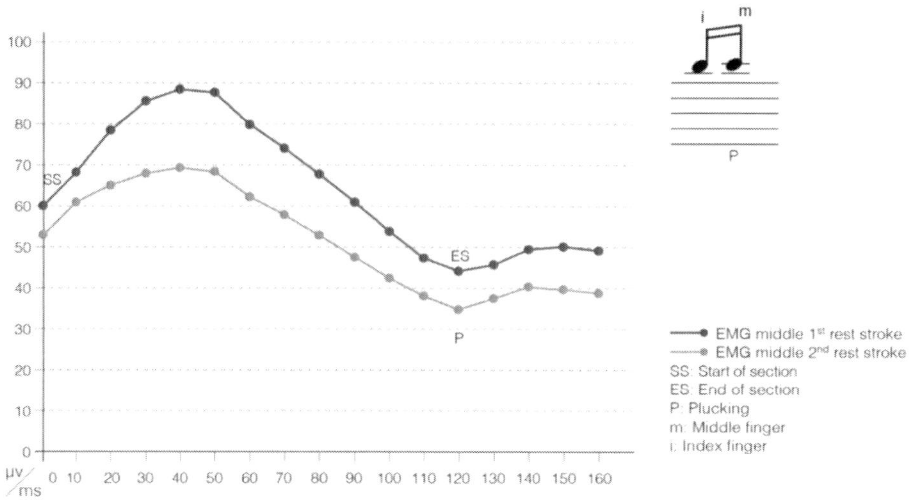


Figure 5. Surface EMG pattern for the middle finger of the right hand, rest stroke, measure 3, second part.

The explanation for why voltage levels increase at the beginning of the action of the middle finger may be that at this moment an impulse must be generated to ensure that the finger reaches the corresponding string and presses it with the required force while counteracting the action of the finger extensors in the opposite direction.

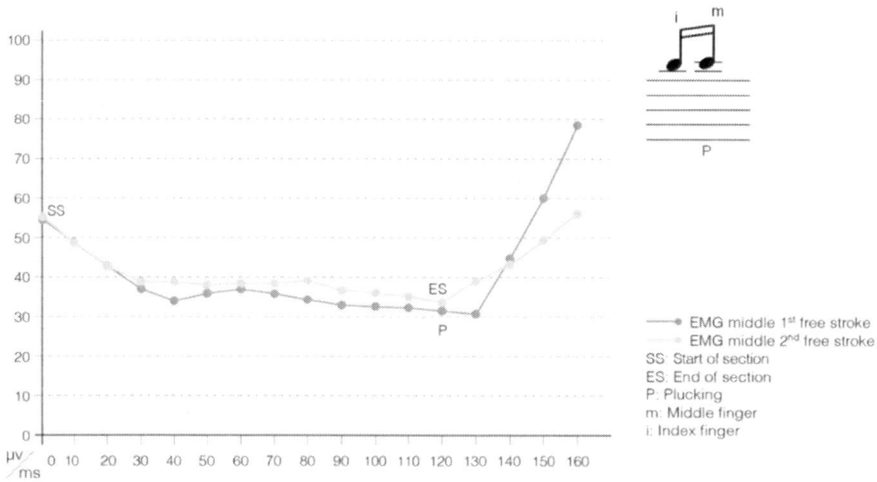


Figure 6. Surface EMG pattern for the middle finger of the right hand, free stroke, measure 3, second part.

As illustrated in Figure 6, when the free stroke was, the level of activation (voltage) dropped at the moment of plucking and increased at the end of the action, i.e., when the middle finger had finished plucking.

The reason why the maximum level activation (voltage) is reached when the string has been plucked (graph value 17), is due to adjustments that may be related to a choice of musical interpretation - something the interpreter decided to apply when performing an ascending scale.

Lastly, the temporal stability presented by Participant 1 in all of the musical passages analysed using both fingers and techniques should be noted.

This provides a clear example of the degree of precision and control that expert performers employ when interpreting a piece.

## Participant 2

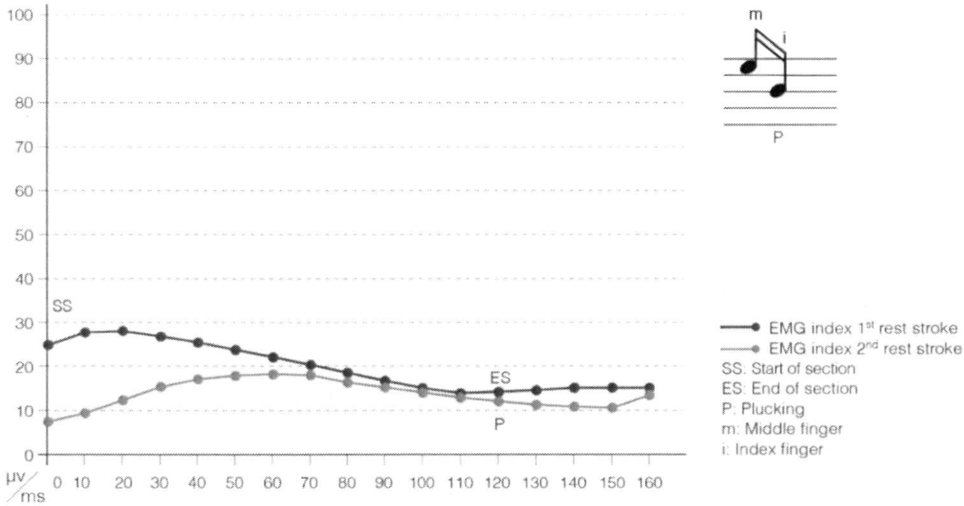


Figure 7. Surface EMG pattern for the index finger of the right hand, rest stroke, measure 6, first part.

In Figure 7, during the execution of the technical section using the rest stroke technique, the participant (guitarist 2) changes from the first string plucked with the middle finger to the second which is plucked with the index finger.

This activation pattern, which lasts from the start of the action until the time of the stroke, can be explained by the need to adjust finger action to control intensity in the execution of these notes.

In Figure 8, when the guitarist plays the passage using the rest stroke technique, the decrease in muscle voltage as the finger approaches the strings to be plucked and during plucking may be motivated by the desire to ensure greater motor control over the execution of the plucking, and to properly adjust this to the intensity with which the corresponding note is to be executed.

As shown in Figure 9, when the guitarist changed from plucking the second string with the index finger to plucking the third string with the middle finger, an increase in the level of muscular voltage was registered during the stroke. This may be due to an attempt to produce a specifically desired tone.

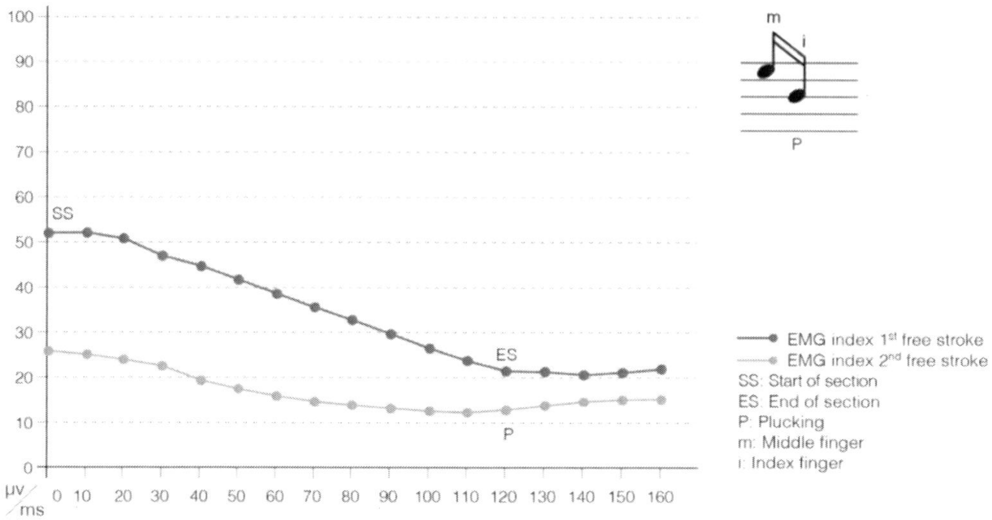


Figure 8. Surface EMG pattern for the index finger of the right hand, free stroke, measure 6, first part.

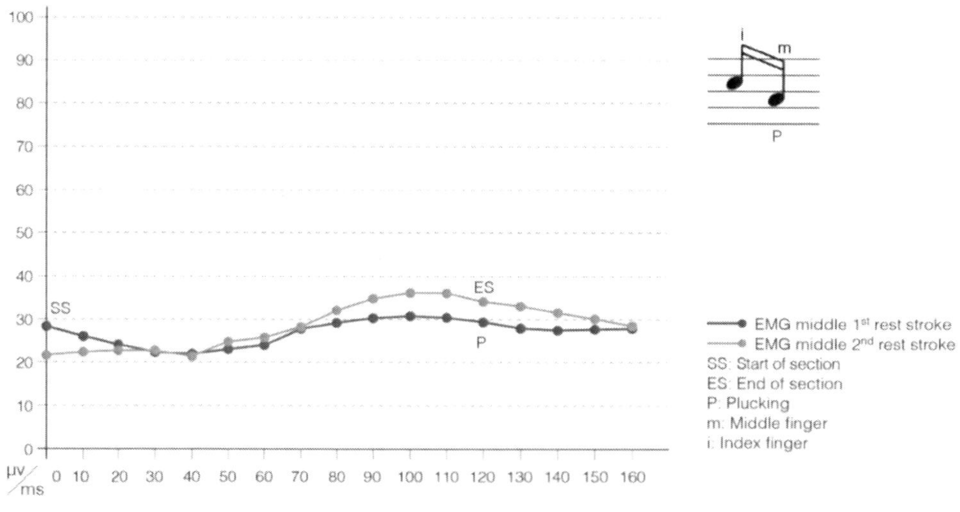


Figure 9. Surface EMG pattern for the middle finger of the right hand, rest stroke, measure 6, second part.

Also, when changing the fingering of the right hand required for the interpretation of the score (index and middle fingers), a situation arises in which the fingering is reversed (with the index finger on the second string and the middle finger on the first string). It is important to note that when playing guitar scales, the index and middle fingers are used primarily, but the ring finger and/or thumb are also used to avoid awkward string crossings Wen-Tzu (1999). The graph shows that this issue does not affect the EMG voltage pattern. The levels remain constant in the two repetitions in the most important moments, which correspond to contact with the string, plucking, and the end of the action (graph values 9-17).

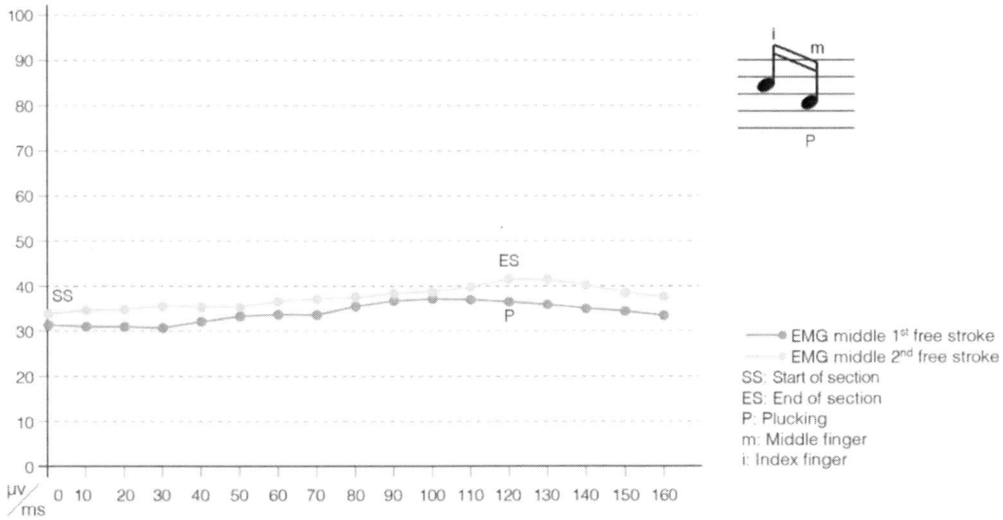


Figure 10. Surface EMG pattern for the middle finger of the right hand, free stroke, measure 6, second part.

In Figure 10 (as in Figure 9 with the rest stroke technique), when the guitarist performed the musical passage using the rest stroke technique, the highest values were recorded at the moment of the stroke. This is comparable to other situations using the index finger in both techniques mentioned above. This may be due to anatomical-functional limitations which intervene in the action of the middle finger, i.e., the middle finger is not completely independent like the index finger because when it is free next to the index finger the intertendinous connection joins it with the ring finger (Latarjet and Ruiz Liard, 1990). These higher values may also be the result of adjustments made by the performer meant to facilitate the action of the finger. In this situation, the same reverse fingering mentioned above takes place. This added challenge did not affect the surface EMG pattern.

Lastly, Interpreter 2 presented similar temporal parameters in all of the repetitions and situations analysed, demonstrating a clear mastery of musical execution.

## CONCLUSION

The results of this study indicated a high degree of stability and consistency in terms of muscle requirements for both participants, which is evidence of the excellent motor control required to become an expert guitarist. It is important to note that each of the performers presents different surface EMG patterns in the fingers (index and middle) and for the stroke systems studied (rest stroke and free stroke), but inter-subject trials were similar for all of the musical passages.

The data showed that Participant 1 presented the same index finger pattern at the moment of the stroke and afterwards in both techniques. It also showed an identical pattern for the middle finger when approaching the strings, during the attack, and in the final stages of the actions.

Participant 2 presented a single pattern for the index finger and a different pattern for the middle finger on the approach to the strings, during the stroke, and at the end of the action.

Of particular importance is the fact that sometimes the participants presented variable muscle activation values in the initial and end phases of their actions for each of the repetitions studied in all musical situations, but there was clear pattern of similar values – which at times were even identical – in key moments of technical difficulty such as contacting and plucking the strings. This means that muscle activation during plucking is directly related to the intensity of execution that the guitarists chose to employ for the interpretation of each note.

Lastly, it should be noted that both performers maintained a high degree of temporal stability in the different musical passages, which demonstrates the extremely high degree of accuracy applied to the interpretation of the musical piece.

In summary, the current study provides an empirical basis to help understand underlying motor control processes in classical guitar performance technique. Further it establishes a process for exploring instrumental performance technique that may be understood as arising from performers` individualized musical motivations or approaches to performing on the guitar.

## References


Basmajian, J. V., & Blumenstein, R. (1989). Electrode Placement in electromyographic biofeedback. In J. V. Basmajian (Ed). *Biofeedback: principles and practice for clinicians* (pp. 378-379). Baltimore: Williams and Wilkins.

Bejjani, F., Ferrara, L., Xu, N., Tomaino, C., Pavlidis, L., Wu, J., & Dommerholt, J. (1989).

Comparison of three piano techniques as an implementation of a proposed experimental design. *Medical Problems of Performing Artists*, *4* (3), 109-103.

Carlevaro, A. (1979). *Escuela de guitarra. Exposición de la teoria instrumental* [*Guitar school Exhibition of instrumental theory*]. Buenos Aires: Barry.

De Contreras, A. (1998). *La técnica de David Russell en 165 consejos* [*David Russell's technique in 165 tips*]. Sevilla: Cuadernos Abolays.

Deschaussées, M. (2002). *El intérprete y la música* [*The interpreter and the music*]. Madrid: Ediciones Rialp.

Duncan, C. (1980). *The art of classical guitar playing*. New Jersey, USA: Summy-Birchard Music.

Feldman, A. G. (1998). Spatial frames of reference for motor control. In M. L. Latash (Ed.), *Progress in Motor Control* (pp. 289-313). Champaign, IL: Human Kinetics.

Fjellman-Wiklund, A., Grip, H., Karlsson, J. S., & Sundelin, G. (2004). EMG trapezius muscle activity pattern in string players: Part I- Is the variability in the playing technique? *International Journal of Industrial Ergonomics, 33,* 347-356.

Granda, J., Barbero, J. C., & Diaz, N. (2006). Diferencias en la actividad de EMG de superficie de los flexores extrinsecos de los dedos en la interpretación de una pieza musical con la guitarra clásica [Differences in surface EMG activity of the extrinsic flexor digitorum in the interpretation of a piece of music with the classical guitar]. En J. Granda, J. (Dir.). *Control Motor: Perspectivas actuales y futuras* (pp. 117-126). Melilla: Ed. Granda.

Granda, J., Barbero, J. C., & Diaz, N. (2007). Aplicación de los principios del aprendizaje motor al estudio de los niveles de calidad interpretativa de una pieza musical con diferentes técnicas en la guitarra clásica [Application of the principles of motor learning to the study of the levels of interpretive quality of a piece of music with different techniques in classical guitar]. *Publicaciones, 37*, 55-72.

Granda, J., Barbero, J. C., & Rodriguez, T. (2004). Las limitaciones informacionales como variables mediadoras en el aprendizaje de instrumentos musicales: un estudio con el piano [The informational limitations as mediating variables in the learning of musical instruments: a study with the piano]. *Motricidad: European Journal of Movement, 11*, 73-88.

Granda, J., Barbero, J. C., & Rodriguez, E. (2010). La técnica instrumental en música: repetición o variabilidad del movimiento. Análisis de la acción del brazo derecho en chelistas expertos [The instrumental technique in music: repetition or variability of movement. Analysis of the action of the right arm in expert cellists]. En R. Reina (ed.), M. Gomis y S. R. Reinoso. *Book de resúmenes del VI Congreso Internacional de la Asociación Española de Ciencias del Deporte*. Elche: Universidad de Elche.

Latarjet, M., & Ruiz Liard, A. (1990). *Anatamia Humana*. México D.F.: Editorial médica panamericana.

Li, Z.M., Zatsiorsky, V. M., & Latash, M.L. (2000). Contribution of the extrinsic and intrinsic hand muscles to the moments in finger joints. *Clin Biomech, 15*, 203-211.

Marques, D. N., Rosset-LLobet, J., Fonseca Marques, M. F., Gurgel, I. G., & Augusto, L. G. S. (2003). Flamenco guitar as a risk factor for overuse syndrome. *Medical Problems of Performing Artists, 18*, 11-14.

Montes, R., Bedmar, M., & Sol Martin, M. (1993). EMG biofeedback of the abductor pollicis brevis in piano performance. *Biofeedback and Self-Regulation, 18* (2), 67-77.

Moore, G. P. (1992). Piano trills. *Music Perception, 9 (*3), 351-360.

Moore, G., Hary, D., & Naill, R. (1988). Trills: Some initial observations. *Psychomusicology, 7* (2), 153-162.

Pujol, E. (1956). *Escuela razonada de la guitarra. Libro 1º* [*Reasoned school of the guitar. Book 1*]. Buenos Aires: Ricordi.

Russell, D. (1993). La técnica clásica [The classical technique]. En M. Stimpson (Ed.). *La guitarra, una guia para estudiantes y profesores* [*The guitar, a guide for students and teachers*] (pp. 181-197). Madrid: Ediciones Rialp.

Schneider, J. (1985). *The Contemporary guitar*. California: University of California Press.

Shan, G., & Visentin, P. (2005). Utility of EMG to quantify activity levels in small muscles. *Medical Problems of Performing Artists, 20* (2), 111-112.

Tárrega, F. (1993). *25 Estudios originales para guitarra, Vol. II* [*Original studies for guitar, Vol. II*]. Valencia: Soneto.

Visentin, P., & Shan, G. (2011). Aplications of EMG Pertaining to Music Performance- A Review. *ArtsBiomechanics*. New York: NOVA Publishers, *1* (1), 15-32.

Wen-Tzu, L. (1999). *A comparative study of the guitar and the Chinese lute-pipa: an overview of their origins, construction and techniques*. Muncie, Indiana: Ball State University.

In: Advances in Arts Biomechanics
Editors: Gongbing Shan and Peter Visentin ISBN: 978-1-53615-641-6


*Chapter 3*

# IMPROVING MUSICIANS' OCCUPATIONAL HEALTH: A PROPOSAL FOR NEW PERSPECTIVES

***Serap Bastepe-Gray***
Johns Hopkins University, Baltimore, MD, US

## ABSTRACT

Despite heightened awareness and interest among both musicians and health-care practitioners within the last twenty years, continuing high rates of new, chronic and repetitive playing related musculoskeletal and neurological disorders in musicians prevail without a significant downward trend. Why? The purpose of this article is to address the knowledge gap at the musician-instrument interface in an effort bring an added perspective into the investigation of playing related disorders in musicians.

## WHAT DOES THE EPIDEMIOLOGICAL DATA INDICATE? DOES IT HELP IN THE DESIGN OF PREVENTIVE AND TREATMENT APPROACHES?

When Dr. Christine Zaza coined the term "playing related musculoskeletal disorders" (PRMD) in the 90s, the scope and the specific manifestations of musicians' playing-related health issues were largely unknown in the health care community. This functional umbrella term with high sensitivity and low specificity was a historical necessity to increase awareness and recognition of this issue, and served to ignite curiosity and research in the field. However, despite an impressive number of studies, considerable uncertainty exists about the nature and etiology of PRMD, its playing-relatedness and non-playing-relatedness, its diagnostic criteria, and outcomes of preventive and rehabilitative approaches. This is in part due to the multifactorial nature of these disorders and a knowledge gap in pathophysiological mechanisms. Additionally methodological challenges in studies such as lack of specific case definitions and gold standards for clinical diagnosis; lack of tools and methods to objectively measure exposure to playing-related musculoskeletal loads; and difficulties in controlling for confounders continue to be a factor in limiting our progress in understanding the magnitude and the nature of these problems, making it difficult to design and assess the effectiveness of prevention and treatment interventions. A look at the body of research on PRMD reveals that the majority of the literature consists of cross sectional data obtained via self-report tools. In most cases these tools do not include criteria with sufficient specificity to inform playing-relatedness or the nature of these complaints. Consequently, the twenty years of research on PRMD informs us that there is a high percentage of musicians likely to be experiencing pain at any given time, but it does not provide us with sufficient evidence on which we can base preventive and treatment approaches.

Among these methodological challenges, defining a set of evidence and consensus based criteria for case definitions of specific and non-specific disorders that fall under the PRMD umbrella is achievable in the short term. It is quite clear that at the surveillance level these criteria need to have enough sensitivity to capture most, if not all, musicians with playing related complaints. However, when an issue is identified, action must follow, as the purpose of active surveillance is prevention. It is difficult to take effective action without understanding the relationship between playing an instrument and the mechanisms that lead to PRMD. An understanding of this requires more specific criteria to begin an explication of exposure-response patterns. Further, the criteria both for surveillance and in the clinical settings need to both assess playing-relatedness of the issues and be able to discriminate between transient pain and symptoms suggestive of disorders. In 2001, Sluiter et al. recognizing a similar need in better identification and classification of work-related upper extremity musculoskeletal disorders in various industry workers in Europe, defined a conceptual model, and established criteria for commonly encountered upper extremity disorders in the worker population (Sluiter, Rest & Frigs-Dresen, 2001). Under the supervision of an international group of experts in occupational medicine, ergonomics, exposure and epidemiology, I propose a similar action from the performing arts medicine community so that our future research on PRMD can generate new knowledge that can translate into clinical and work-place prevention and treatment approaches to musicians' PRMD.

## Prevention versus Rehabilitation

Most of our efforts within the last twenty years in helping musicians combat PRMD has focused on primary prevention, with increasingly more vocalized demands at various organizational levels on instrumental instructors to integrate health and wellness training in their studio teaching. However, a similar demand within the health care system for improved diagnostic criteria, standards of treatment and evidence-based outcome monitoring has been lacking. Despite evidence that supports integration of biomechanical and ergonomic approaches into preventive and treatment approaches to decrease time away from work and to provide effective long-term return-to-work in other industries, approaches to musicians in the clinical rehabilitation setting is still largely based on the traditional PMSS paradigm: Physical agent modalities (heat packs, cold packs, ultrasound etc.), Manual therapy and massage, Strengthening and Stretching. The literature is limited on outcomes reporting for PRMD treatment approaches, so we do not know how effective these methods are in bringing a relief to the musicians suffering from PRMD and in helping them return to playing their instruments. Considering the high rates of chronic and repetitive PRMD reported, even if these intervention methods can promote tissue healing, it is reasonable to conclude that the musician is destined to return to the biomechanical patterns that induced the problem in the first place unless the treatment approaches target modification of motor behaviors on the instrument. This is no easy feat: musicians' work involves non-normative use of upper extremities and is driven by the sonic outcomes that are achieved by minute, precise and fast movements that happen at the interface of the sound producing mechanisms of their instrument and the trained facility of their fingers. Due to the sensory-motor precision required in playing an instrument, unless a health care practitioner or an ergonomist has internalized sensory-motor knowledge in playing a similar instrument, it is next to impossible to detect the playing-related biomechanical loads through observation, especially the loads on the distal upper extremity arising from subtle alignment issues, lack of micro-unloading habits and excessive force production by the fingers.

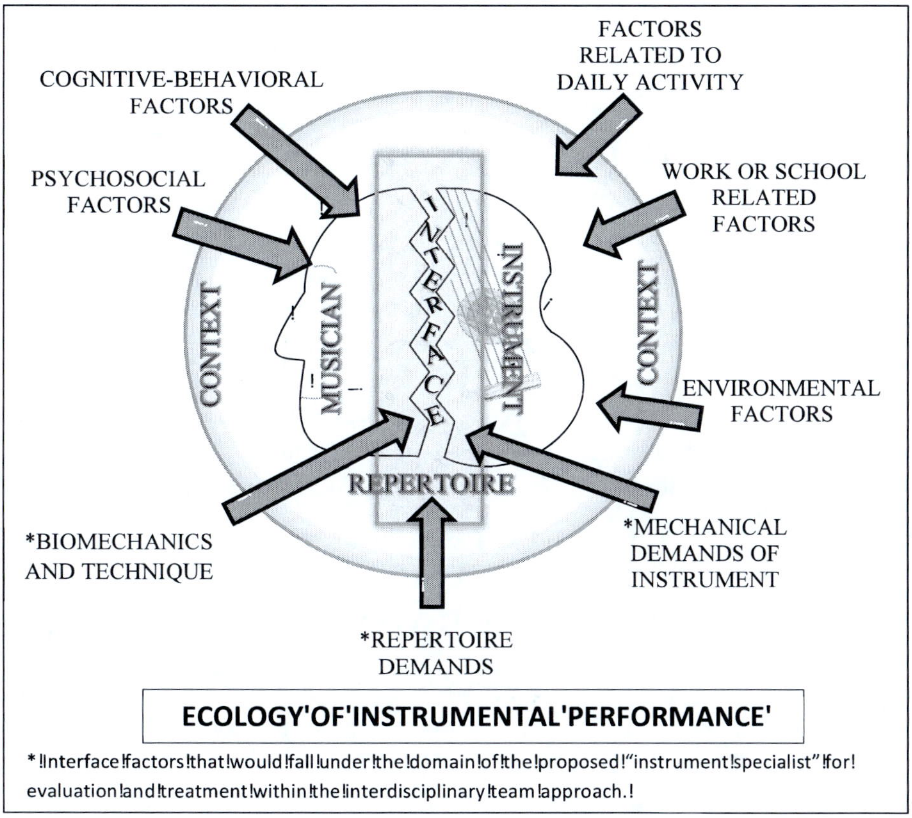


Figure 1. The musician-instrument-repertoire complex and the factors that influence the work of musicians.

Currently, one of the major obstacles in the field is lack of tools and methods that can objectively measure these loads with high utility in the musicians' work-place and in the clinical setting. The small number of rehabilitation specialists with interest in musicians' PRMD, consequently, find themselves in an evaluative and treatment capacity limited to the proximal and larger joints of the upper extremities – modifications of which may or may not be feasible without a thorough understanding of the upper extremity biomechanics of playing a particular family of instruments, as these larger joints are part of the kinematic chain where the action is happening at the musician-instrument interface. Any modification of positions and movements of these larger joints will have a significant impact on this interface. Additionally, the value of postural correction of the trunk and proximal upper extremity without integrating a plan to decrease the loads arising from the subtle misalignments at the wrist and finger joints, as well as from sustained and peak finger-force production while preserving the precision and accuracy necessary to play an instrument is questionable in improving long term outcomes of prevention and treatment approaches.

Until the research and development of such tools with validity, reliability and utility with which we can arm our physicians, therapists and ergonomists, I propose that the performing arts medicine and musician communities coordinate their respective areas of expertise to develop an "instrument specialist" position, parallel to a "singing-voice specialist" that acts as a liaison between the fields for the benefit of the musician with PRMD. Ideally an instrument specialist would be a musician with professional performance training on an instrument, be knowledgeable in performance on the family of instruments of his/her own instrument, have experience in instruction on that instrument and have a working knowledge of functional anatomy and upper extremity kinesiology. The role of the instrument specialist within the interdisciplinary team would focus on interventions targeted to the complex interface of musician-instrument-repertoire within an ecological framework (Figure 1) (Bastepe-Gray, 2014).

Position papers from professional organizations such as PAMA, MTNA and ASTA that emphasize the importance of bringing the expertise that resides in the informed instrumental pedagogue into the clinical setting would help energize the efforts for the development of a training and certification program for musicians interested in acting in this role. In the long run, a strong action at the organizational level might even effect change in the care delivery model and third party re-imbursement systems to include instrument specialists as evaluation and treatment consultants with billable services, thus reducing the financial barrier between musicians and specialized care.

## How about PRMD Management?

Considering the pervasiveness of the problem and the debilitating impact on the mental and physical health of musicians, perhaps a management approach to PRMD, akin to a disaster management approach modeled by Federal Emergency Management Agency (FEMA), that includes interventions to ensure preparedness, response, recovery and mitigation of the issue, would yield better outcomes in decreasing PRMD prevalence in musicians. Preparedness, or "a state of readiness to respond to a disaster, crisis or any other type of emergency situation" as defined in the field of emergency management, could translate into PRMD management as training for early awareness of the problem and injury first-aid, contact with the treatment team prior to the episode of PRMD, and at the organizational level, establishing pipelines that decrease barriers for early contact between the musician and the treatment team at the first indication of potential PRMD in an effort to ensure effective response. Response activities would include timely administration of injury first-aid (PRICE/MICE) protocols followed by an early clinical visit for diagnosis and treatment by an interdisciplinary team under the leadership of a board-certified physician, practice/performance and work schedule modification, modification of daily living activities and rehabilitation counseling. Recovery would need to consider not just tissue healing through therapy, but also the recovery of the various life and musician roles and activity patterns of the musician. Mitigation would require identifying risks and developing long-term strategies to reduce or eliminate long term risks associated with the development of another PRMD episode.

## CONCLUSION

In summary, the next stage of efforts in decreasing PRMD rates in musicians would highly benefit from:

1. Clarifying case definitions of specific and non-specific disorders that fall under the umbrella of PRMD,
2. Elucidating the role of specific playing-related biomechanical loads in the etiopathogenesis of these disorders,
3. Integrating retraining into rehabilitation through acknowledging and clarifying the local areas of expertise of the health care professional and the instrumental specialist – neither one is complete in itself,
4. Contextualizing the instrumental practice/performance within the daily living activities to better manage daily work-loads on upper limbs,
5. Developing intervention frameworks that include PRMD management in addition to PRMD prevention.

## REFERENCES

Bastepe-Gray SE. (2014). Healthy hands: A primer in injury management for guitarists. *Soundboard.* 40(2):24-32.

Sluiter JK, Rest KM, Frigs-Dresen M. (2001). Criteria document for evaluating the work-relatedness of upper-extremity musculoskeletal disorders. *Scandinavian Journal of Work & Environmental Health* [*Supplement*]. 27(1):1-102.

In: Advances in Arts Biomechanics
Editors: Gongbing Shan and Peter Visentin

ISBN: 978-1-53615-641-6


*Chapter 4*

# FEASIBILITY OF FINE WIRE ELECTROMYOGRAPHY FOR ORCHESTRAL VIOLA TASKS

***Leila K. Kelleher[1], Ryan J. Frayne[2], Timothy J. Doherty[3] and James P. Dickey[3]***
[1]Humber College, Canada
[2]Dalhousie University, Canada
[3]Western University, Canada

## ABSTRACT

*Background:* Professional orchestral musicians are at high risk of injury due to the volume and repetitive nature of their playing. Previous research has mainly investigated the larger muscles of the upper extremities using surface EMG, however, the activation patterns of forearm and hand muscles have not been investigated. Since small muscles should be analysed using fine-wire EMG to prevent cross-talk between muscles, it is important to determine if this methodology is practical when studying musicians. Furthermore, earlier studies have typically focused on low effort tasks, which likely underestimate the muscular demands of the professional orchestral workplace.

*Purpose:* To determine the utility of fine-wire EMG as a method for describing muscle activation of viola players while performing professional-level playing tasks.

*Methods:* Four professional viola players (age 41 ± 10.4; years performing professionally 21 ± 8.3) performed a variety of orchestral excerpts from the standard repertoire. Muscle activation was measured and data collected using surface and fine-wire EMG on the right upper limb.

*Conclusion:* This study confirms the feasibility of using fine-wire EMG to evaluate the activation of small muscles in the forearm during musical performance. Muscle activations were higher for some orchestral excerpts than for other, less strenuous tasks. Because of this, we recommend that future research should consider using a variety of professional-level excerpts of varying complexity to ensure that resulting data is indicative of typical workplace exposures.

**Keywords:** viola, biomechanics, musicians, musicians' injuries

## INTRODUCTION

Orchestral musicians exhibit high rates of occupational injury with a reported lifetime prevalence as high as 84% (Ackermann, Driscoll, & Kenny, 2012; Jonsson, 1988) and a point prevalence between 39-87% (Zaza, 1998). This is similar to prevalence rates for some occupations, such as female supermarket checkers (50% (Margolis & Kraus, 1987)), but is considerably higher than those found in other professions that are generally perceived to have high risk rates, such as male steel workers (25.5% (Choi, Kang, Kim, & Han, 2009)). Orchestral musicians may spend up to 22.5 hours a week rehearsing and performing in addition to personal practice time (Guptill, 2011), which is thought to be related to the high injury rates observed (Bird, 2013).

Since many of these injuries are due to overuse (Fry, 1986), several studies have examined muscle activation patterns in string musicians. Studies using electromyography (EMG) to measure muscle activation have typically focused on the larger muscles of the upper limb, primarily the biceps, triceps, deltoids, and trapezius. One study reported biceps and flexor carpi radialis muscle activations of approximately 10% (Shan, Visentin, & Schultz, 2004), however these low levels of activation do not appear to be consistent with the high rates of injury as they do not exceed endurance limits (fatigue threshold of 14% MVC) for intermittent static contractions (Björkstén & Jonsson, 1977). Another study which examined muscles of the shoulder girdle using surface and fine-wire EMG reported supraspinatus to be the only muscle that exceeded the fatigue threshold during a bowing exercise task (Rickert, Halaki, Ginn, Barrett, & Ackermann, 2013). Much of the research describing muscle activation during string instrument playing has used protocols involving basic performance tasks, such as scales and bowing exercises. These do not represent the typical technical demands made of professional players (Bejjani, Ferrara, & Pavlidis, 1990; Rabuffetti, Converti, Boccardi, & Ferrarin, 2007; Rickert, et al., 2013; Shan, et al., 2004).

When using surface EMG, cross-talk, or signal interference from adjacent or superficial muscles can be a significant problem, especially when targeting smaller muscles, because the detection area of the electrodes can be larger than the belly of the target muscle or the muscle may be deep (J. V. Basmajian & De Luca, 1979). This has been identified as a specific problem in one study collecting muscle activation data for small muscles in a trumpet player's embouchure (Shan, Visentin, & Iltis, 2005). Fine-wire EMG (fwEMG), which eliminates cross-talk problems and allows access to deep muscles, seems a logical solution for muscle activation studies. However there are concerns that fwEMG is invasive and may affect motor control during performance (Visentin & Shan, 2011). To date, only one study has successfully used fwEMG in a case study of a single cellist for deep muscles of the shoulder girdle (Rickert, et al., 2013).

Given the low level of muscle activation reported in previous studies and the high prevalence of musculoskeletal injuries, we wished to determine if the muscular demands of professional viola players are greater than previously reported. In addition to measuring activation using surface EMG of the large muscles of the right upper limb, we were interested in evaluating the activations of the smaller muscles of the forearm and hand using fwEMG to determine if it is an appropriate methodology to use in musicians.

# METHODS

Four professional viola players (age 41 ± 10.4 years; performing professionally 21 ± 8.3 years) participated in this study. All subjects (S1, S2, S3, S4) were in a good state of performing health, with none having experienced injury for two years prior to testing. The study protocol was approved by the institutional REB and all subjects provided written, informed consent. Basmajian-style single-needle fine-wire electrodes (J. Basmajian, 1974) were manufactured and autoclaved within the lab. Each electrode consisted of two 100 μm diameter insulated stainless steel wires (California Fine Wire, Grover City, California) with approximately 5 mm of insulation removed from the tip.

The electrodes were inserted into four muscles of the subjects' right upper extremity (first dorsal interosseous - FDI, flexor carpi radialis - FCR, extensor carpi radialis brevis - ECRB, flexor digitorum superficialis - FDS) using landmarking techniques described in the literature (Perotto, 2011). The electrodes were placed using a hypodermic needle (25G x 25mm; BD, Franklin Lakes, New Jersey) as a cannula (J. Basmajian, 1974). Each muscle was contracted to ensure that the wires were secured within the muscle, and the hypodermic needle was withdrawn leaving the fine-wire electrodes within the muscle belly. Surface electrodes were also placed over the belly of the biceps (B), mid-deltoid (D), and brachioradialis (BR). As recommended, each target muscle was contracted and the EMG was examined to ensure that the electrodes were placed in/on the appropriate muscle belly (J. V. Basmajian & De Luca, 1979). EMG signals were transmitted wirelessly from the subjects to the Trigno base system (Delsys Inc, Boston, MA, USA) at a sampling rate of 2000 Hz using a 16 bit A/D converter (National Instruments Inc., Austin, TX, USA) and recorded using a custom LabVIEW program (version 12.0, National Instruments Inc., Austin, TX, USA).

A metronome (BOSS Dr. Beat DB-90, Roland Corp., Los Angeles, CA, USA) was used throughout the testing to ensure standardized tempi. A scale was used to establish a baseline for comparison and several standard orchestral excerpts were performed. The scale was performed at 80 beats per minute. Excerpts were chosen to explore a variety of bowing techniques, dynamic levels and tempi, typical of orchestral repertoire. Details of the excerpts are presented in Table 1.

All excerpts were performed with a typical bow hold except for Roman Carnival pizzicato and Capriccio Italienne pizzicato excerpts, which were performed pizzicato (strings are plucked using the right index finger) with bow held using a palmar grip. All of the subjects performed the excerpts in the same order, as randomization with this small number of subjects would have had no utility.

All subjects were videotaped during their performance, and a sync light was used to synchronize EMG and video data. Each excerpt was identified using the sync signal at the beginning of playing time and the end of the excerpt was determined using video data. EMG data were high-pass filtered at 30 Hz, full-wave rectified, and low-pass filtered with a second-order Butterworth filter at 6 Hz, as recommended by others (Shiavi, Frigo, & Pedotti, 1998; Winter, 1990), using a custom LabVIEW program (National Instruments Inc., Austin, TX, USA). All trials of orchestral excerpts were normalized to the mean activations from the baseline scale trial. This approach to baseline normalization is commonly used in gait analysis (Yang & Winter, 1984).

**Table 1. Details of excerpts used in testing**

| Composer | Work | Excerpt | Tempo | Key Technique |
|---|---|---|---|---|
| - | C major scale | 2 octave scale | ♩ = 80 bpm | *Legato* (smooth, connected bowing) |
| Berlioz | *Roman Carnival* Overture | 2 bars after rehearsal 1 to rehearsal 3 | ♩ = 52 bpm | *Legato* |
| Mendelssohn | *Scherzo* from a Midsummer Night's Dream | Rehearsal C to D | ♩. = 85 bpm | *Spiccato* (off-the-string bow stroke) |
| Tchaikovsky | *Capriccio Italienne* | Measure 180 to bar after D | ♩ = 120 bpm | *Ricochet* ("thrown" bow stroke) |
| Holst | *Mars* from The Planets | Beginning to measure 9 | ♩ = 136 bpm | *Col legno* (wood of the bow tapped on the string) |
| Berlioz | *Roman Carnival* Overture | Measures 21 to 24 | ♩ = 52 bpm | *Pizzicato* (plucked strings) |
| Tchaikovsky | *Carpiccio Italienne* | Measure 148 to 156 | ♩. = 62 bpm | *Pizzicato* |
| R. Strauss | *Don Juan* | Beginning to measure 36 | ♩ = 156 bpm | *Martelé* (heavy, detached bow stroke) |
| Beethoven | Symphony no. 6 *"Pastoral,"* mvt IV *(Gewitter sturm)* | Rehearsal E to measure 89 | ♩ = 184 bpm | *Martelé* |
| Mendelssohn | *Scherzo* from a Midsummer Night's Dream | Rehearsal C to D | ♩. = 85 bpm | *Spiccato* |
| - | C major scale | 2 octave scale | ♩ = 80 bpm | *Legato* |

## RESULTS

A high degree of variability in muscle activation between subjects was observed. The general activation patterns for S1 and S2 were similar. S4 demonstrated higher activations, particularly in the FCR, than the other subjects. S3 often demonstrated lower activations than the other subjects. Mean muscle activation data, normalized to the baseline scale, is presented in Figure 1.

S1, S2 and S4 exhibited higher activations in all muscles in the Beethoven excerpt compared to the baseline scale, although the activation in ECRB was only slightly increased (104% of the baseline). S1 exhibited extremely high peak levels of activation for the BR muscle in this excerpt (550% of the baseline). FCR was also more highly activated in this excerpt (333% 200%, 500% of the baseline) for S1, S2 and S4, respectively.

FCR activations were higher than for the baseline scale in: the Roman Carnival, Mendelssohn Scherzo, Capriccio Italienne ricochet, Capriccio Italienne pizzicato, Don Juan, and Beethoven, excerpts. They were only lower than the baseline scale in the col legno and Roman Carnival pizzicato excerpts.

In most excerpts, subjects activated their FDIs at a similar level to the baseline scale. The highest activation in this muscle was in the Beethoven excerpt. Activations of the ECR were all lower than the baseline scale, except for Beethoven, which was 104%. S4's activations were less than 100% of the baseline scale in all trials for this muscle. All subjects' FDS muscle activations were similar to the baseline scale for most excerpts, except for the Roman Carnival pizzicato excerpt, which had considerably lower activation levels, and the Don Juan and Beethoven excerpts, which had higher levels of activation. The BR of our subjects were activated at a similar level as the baseline scale during the quiet and moderately loud excerpts. In contrast, the louder excerpts (Capriccio Italienne pizzicato, Don Juan, and Beethoven), elicited higher activations.

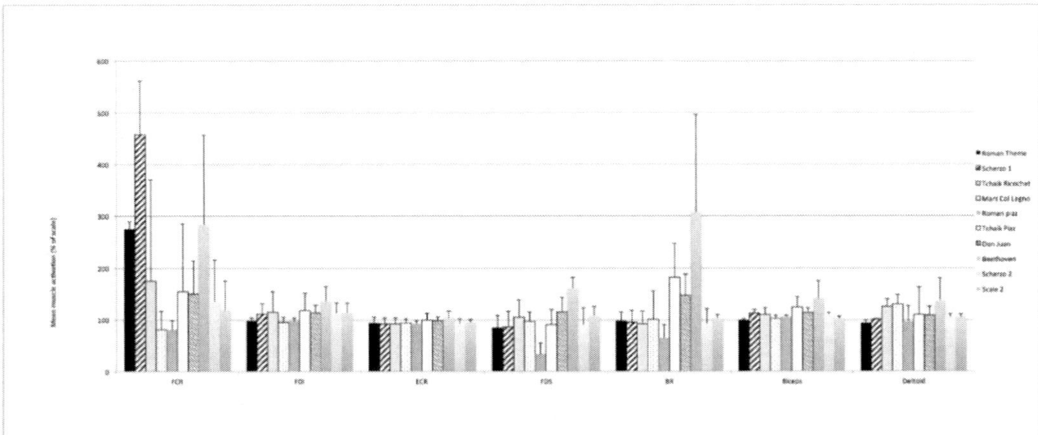

*The FCR in the Roman pizz, Roman Theme, and Scherzo 1 do not include subject 1 due to technical difficulties. **The mean for BR in Tchaik Ricochet does not include subject 1 due to technical difficulties.

Figure 1. Mean muscle activation* normalized to scale task. Each profile showed distinct profiles of muscle activation for the different muscles.

Overall, our subjects' biceps and deltoid muscles were activated at similar levels as the baseline scale, except for the deltoid in col legno and Beethoven, and the biceps in Capriccio Italienne pizzicato and Beethoven.

All subjects showed similar or lower activation than the baseline scale for all muscles during both Roman Carnival excerpts. For the FDS, activations in all subjects were lower than the baseline scale in the Roman Carnival pizzicato excerpt (22%, 12%, 44%, 56% of the baseline). The Col Legno technique exhibited variable FCR activations of 66%, 50%, 75%, and 133% of the baseline.. Similar variability between subjects was seen for the FCR in Mendelssohn excerpt (133%, 83%, 75%, 250% of the baseline), in the FDS in Capriccio Italienne pizzicato (97%, 52%, 122, 92% of the baseline), and the FDS during the Capriccio Italienne ricochet excerpt (121%, 71%, 144%, 81% of the baseline).

## DISCUSSION

Previous studies have reported that the large upper extremity muscles are activated at relatively low levels while performing scales and bowing exercises; they report that the average muscle activity was less than 20% MVC, with some muscles consistently less than 5% MVC (Shan, et al., 2004; Wales, 2007). In contrast, our results show higher muscle activations during performance of strenuous excerpts when compared to scales, with one subject achieving peak activations of 550% of the baseline scale in the BR during the Beethoven excerpt.

We also observed differences in muscle activation patterns between participants, for instance, subject 4 tended to have higher muscle activations in the FCR compared to the other subjects. We believe that these differences may be explained by variation in bowing technique. This variability in bowing technique is almost certainly due to the oral traditions of string pedagogy (Shan & Visentin, 2010).

In most excerpts, the FCR was more highly activated than the scale. In the case of the Mendelssohn, the mean FCR activation between subjects was 458% of the baseline scale. This higher activation of the wrist flexors may help to explain the high rates of muscle overuse and tendinitis in pre-professional music students (Cayea & Manchester, 1998).

There were distinct differences in muscle activation levels between the Roman Carnival pizzicato and Capriccio Italienne pizzicato excerpts. This difference is most likely indicative of the technique of pizzicato the subjects used. We observed large standard deviations in Capriccio Italienne pizzicato excerpt (as large as 131% of the baseline in the FCR), suggesting that our subjects used different technical strategies to play the very fast pizzicato required.

Two of our subjects complained of pain at the fwEMG insertion sites, which may have inhibited or changed their performance. The earlier fwEMG study did not find that the electrodes interfered with the playing technique of their single subject (Rickert, et al., 2013), so this may be highly dependent on the individual and the specific muscles studied. This discomfort (and potential change in technique/muscle recruitment) reported by half of our subjects confirms concerns in the literature regarding the fwEMG electrodes affecting "outcomes of performance" in musicians (Visentin & Shan, 2011). Strategies for minimizing fwEMG electrode pain could include reinsertion of the electrodes to minimize discomfort (slight adjustment of the insertion site can alleviate pain), or by avoiding insertion points near motor end plates, although this may be a challenge with smaller muscles, such as the FDI.

We experienced significant challenges collecting accurate maximum voluntary contraction (MVC) data for normalization, which is why we chose to use a baseline task for this purpose. The deficiencies in our MVCs were revealed when we normalized louder and more difficult excerpts. We observed trial activations as high as 1700% of MVC, which confirmed that our MVCs were not accurate. We believe this may be due to the difficulty of maximally contracting small muscles and that our subjects were likely inexperienced in maximal muscle contractions. It is important to note that we did not observe MVCs of greater than 100% in our scale trials which means that inaccurate MVCs may not be evident in less technically demanding tasks. We recommend that future researchers consider using a baseline task for EMG normalization in musicians.

## Conclusion

This study confirms that it is feasible to use fine-wire EMG to evaluate the activation of small muscles in the forearm (and, to a lesser extent, the hand) during musical performance, however care must be taken to ensure that the player does not experience pain. Our findings of high muscle activation during orchestral excerpts support established frameworks (such as biomechanical modeling) for evaluating overuse syndromes (Visentin & Shan, 2004) and suggest that the style of repertoire should inform the duration and intensity of orchestral rehearsals and performances in order to reduce the high levels of injury in the orchestral workplace.

Muscle activations were higher in some orchestral excerpts than in less strenuous tasks. Because of this, we recommend that future research use a variety of tasks to ensure that the resulting data is indicative of typical workplace exposures. This approach could be used to inform future injury reduction strategies in the orchestral workplace.

## REFERENCES

Ackermann, B., Driscoll, T., & Kenny, D. T. (2012). Musculoskeletal pain and injury in professional orchestral musicians in Australia. *Medical Problems of Performing Artists*, 27(4), 181.

Basmajian, J. (1974). Simple method to improve performance of fine-wire electrodes. *American Journal of Physical Medicine & Rehabilitation*, 53(6), 269-270.

Basmajian, J. V., & De Luca, C. J. (1979). *Muscles alive: their functions revealed by electromyography*: Williams & Wilkins.

Bejjani, F. J., Ferrara, L., & Pavlidis, L. (1990). *A comparative electromyographic and acoustic analysis of violin vibrato in healthy professional violinists:* New York University.

Bird, H. (2013). Overuse syndrome in musicians. *Clinical rheumatology*, 32(4), 475-479.

Björkstén, M., & Jonsson, B. (1977). Endurance limit of force in long-term intermittent static contractions. *Scandinavian journal of work, environment & health*, 23-27.

Cayea, D., & Manchester, R. A. (1998). Instrument-specific rates of upper-extremity injuries in music students. *Organ*, 26(362), 11-14.

Choi, W.-J., Kang, Y.-J., Kim, J.-Y., & Han, S.-H. (2009). Symptom prevalence of musculoskeletal disorders and the effects of prior acute injury among aging male steelworkers. *Journal of occupational health*, 51(3), 273-282.

Fry, H. J. (1986). Incidence of overuse syndrome in the symphony orchestra. *Med Probl Perform Art*, 1(2), 51-55.

Guptill, C. (2011). *The lived experience of working as a musician with an injury*. Work, 40(3), 269-280.

Jonsson, B. (1988). The static load component in muscle work. *European journal of applied physiology and occupational physiology*, 57(3), 305-310.

Margolis, W., & Kraus, J. F. (1987). The prevalence of carpal tunnel syndrome symptoms in female supermarket checkers. *Journal of Occupational and Environmental Medicine*, 29(12), 953-956.

Perotto, A. O. (2011). *Anatomical guide for the electromyographer: the limbs and trunk*: Charles C Thomas Publisher.

Rabuffetti, M., Converti, R. M., Boccardi, S., & Ferrarin, M. (2007). Tuning of the violin-performer interface: an experimental study about the effects of shoulder rest variations on playing kinematics. *Medical Problems of Performing Artists*, 22(2), 58-66.

Rickert, D. L., Halaki, M., Ginn, K. A., Barrett, M. S., & Ackermann, B. J. (2013). The use of fine-wire EMG to investigate shoulder muscle recruitment patterns during cello bowing:

The results of a pilot study. *Journal of Electromyography and Kinesiology*, 23(6), 1261-1268.

Shan, G., & Visentin, P. (2010). *Arts Biomechanics: An Infant Science: Its Challenges and Future*: Nova Science Publishers.

Shan, G., Visentin, P., & Iltis, P. W. (2005). Utility of EMG to quantify activity levels in small muscles. *Medical Problems of Performing Artists*, 20(2), 111-113.

Shan, G., Visentin, P., & Schultz, A. (2004). Multidimensional signal analysis as a means of better understanding factors associated with repetitive use in violin performance. *Medical Problems of Performing Artists*, 19(3), 129-139.

Shiavi, R., Frigo, C., & Pedotti, A. (1998). Electromyographic signals during gait: criteria for envelope filtering and number of strides. *Medical and Biological Engineering and Computing*, 36(2), 171-178.

Visentin, P., & Shan, G. (2004). An innovative approach to understand overuse injuries: biomechanical modeling as a platform to integrate information obtained from various analytic tools. *Medical Problems of Performing Artists*, 19(2), 90-97.

Visentin, P., & Shan, G. (2011). Applications of EMG Pertaining to Music Performance-A Review. *Arts BioMechanics*, 1(1), 15.

Wales, J. (2007). *3D movement and muscle activity patterns in a violin bowing task.*

Winter, D. A. (1990). *Biomechanics and motor control of human movement*: John Wiley & Sons.

Yang, J. F., & Winter, D. A. (1984). Electromyographic amplitude normalization methods: improving their sensitivity as diagnostic tools in gait analysis. *Archives of Physical Medicine and Rehabilitation*, 65(9), 517-521.

In: Advances in Arts Biomechanics
Editors: Gongbing Shan and Peter Visentin ISBN: 978-1-53615-641-6


*Chapter 5*

# STRESSING "UNSTRESSED:" RELAXED EMPHASIS AND LIGHTNESS IN THE FLOW OF MOVEMENT – UNDERSTANDING METER VIA PHYSICAL ASSISTANCE AND SUPPORTED METRICAL SOUND EXPERIENCE IN BOWED STRING INSTRUMENTAL PERFORMANCE*

***Julia von Hasselbach***
University of Freiburg, Germany

## ABSTRACT

This article is a slightly changed English version of the first part of an article published in the proceedings of the conference "Understanding and Experiencing Music with the Body" ("Musik verstehen und erleben mit dem Körper," Hanse-Wissenschafts-Kolleeg Delmenhorst, Germany 2016). In an interdisciplinary dialogue, this contribution presented aspects of instrumental education, in particular, of biomechanical research in the field of bowed string instrumental performance.

Provided that a biomechanically elaborated concept of bowing 'technique' exists while teaching string instrumental performance, with clear perceptible correlations between movements and the resulting sounds, the method of physical assistance can be used to support, for example, the understanding and the performance of musical meter. With the help of a teacher giving suitable assistance, students can feel the necessary movements in their own body and experience metrical sounds of nearly professional quality arising from their own instruments, without yet knowing how to do it themselves. Motor learning and musical learning are both stimulated through a distinct development

* First published as "Schwere und Leichtigkeit im Fluss der Bewegung – Metrisches Verstehen und Klangerleben im Streichinstrumentalspiel", in L. Oberhaus & Chr. Stange (eds.), *Musik und Körper: Interdisziplinäre Dialoge zum körperlichen Erleben und Verstehen von Musik* (pp. 121-143). Bielefeld, Germany: [transcript] 2017, translated by permission of [transcript] publishers 2017.

of musical objectives and movement objectives, which should finally result in a compelling autonomous performance.

Taking Eichhorn's theory of a leverage system in string instrumental bowing as the technical guiding principle for physical assistance, emphasis is revealed as the result of relaxed swinging movements without the need of physical stress to play a 'stressed' tone. The refined interaction of levers allows for artistically convincing metrical tonal shapes with excellent transitions from emphasis to lightness. Metrical tonal shapes are looked at in theory and praxis, revealing the peculiarities of bowed string instruments. The method of physical assistance is discussed in the context of instrumental education and finally demonstrated in a concrete example of metrical bowing in triple meters on the violin.

## 1. Background and Aims

"Time [...] is the soul of music." (Leopold Mozart)[1]

Music without a noticeable meter is rare, and music that engages the listener physically is especially marked by great metrical composition, be it classical or popular music. For example, in a pop song such as Astronaut[2], that is designed in the supposedly simple $^4/_4$ meter, the composers utilize metrical effects in a sophisticated way: The withdrawal of an otherwise strongly pulsating meter in the bridge may create a breathtaking impression of ethereal floating in listeners, as if they were looking at the earth themselves. The soft resumption of the more strongly pulsating meter enables listeners to feel especially invigorated and possibly activated to change their lives despite the paradoxically pessimistic turn of the lyrics in the transition to the refrain. At the latest, with the return of the weighty refrain, there is no stopping the emotional and physical resonance. Of course, all musical parameters contribute in achieving such gripping effects of music, but the following considerations will focus on musical time in the sense of meter, especially on how to impart knowledge about its physical transformation into sound while teaching string instrumental performance.

At the beginning of his famous "Treatise on the Fundamental Principles of Violin Playing", Leopold Mozart already pointed out the importance of learning metrical principles: "Time [...] is the soul of music. [...] Everything depends on musical time-measure, and the teacher must use the greatest patience in seeing that the pupil grasps it thoroughly, with diligence and attention" (Mozart 1985, p. 30). From a pedagogical perspective, Mozart utilized a method of physical assistance[3] and thereafter independent performance: "For this purpose it will be advisable for the teacher to constantly guide the pupil's hand according to the beat, and [...] allow him to beat time himself [...]" (Mozart 1985, p. 33).

For this conducting approach, it may have been of significance that Leopold Mozart assumedly learned to play the violin mostly autodidactically. It is not known who may have been his violin teacher otherwise. One teacher that could come into question would have been

[1] Mozart 1985, p. 30.

[2] Composed by Andreas Bourani, Simon Müller-Lerch, Paul Neumann, Marek Pompetzki, Cecil Remmler and Paul Würdig (Sido); the song received the Award of German Music Composers for "The Most Successful Work of 2015".

[3] Physical assistance today is perhaps more frequently used within the realm of sports pedagogy to support torques, such as in exercises on the bars, than within string instrumental performance to create pivots and swinging levers.

a friendly cathedral organist, who surely rather conveyed basic musical knowledge than instrumental technique (cf. Mozart 2007, p. 14). Although Leopold Mozart attributes such an important role to metrical education at the start of the first standard work in the history of violin didactics, it seems odd how sparsely methodically he goes into this in the continuation of his work. Moreover, it is astounding that the learning of metrical performance is only implicitly found in contemporarily established violin methods as well. Therefore, it falls to the respective teacher to support the learning of a metrically elaborated performance to the best of his or her knowledge and belief.

Why, especially in the understanding of metrical matters, the teachers and all sorts of acoustically experienceable models are of such importance is evident: The notation offers sparse information. In classical music, pulse-groups are indicated by the time signature, and a disciplined attention to bar lines is basically helpful in metrical orientation, but characteristic metrical properties stop being noted at the latest in Jazz and folk music. A folky backbeat groove (a short stress on each 'and'-time in rapid succession), for instance, is, in my opinion, only perceivable through an auditively and physically comprehensible metrical experience. This is especially important if the stress does not coincide with a separate bow stroke and therefore easily articulatable tone, but is instead briefly realized within a bound sequence, or even within an already sounding tone.

In their more abstract standard forms, conducting beat patterns only resemble the bowing movements in a $^{2}/_{4}$ time signature. Within a $^{4}/_{4}$ time, a light deviance from the 'sterile' 1-stroke on a point straight in front, through a light swinging movement of the upper arm away from the body suffices: The movement becomes immediately almost identical to a bowing movement. This can allow for a sometimes sensible, helpful gesture – not only while working with youth orchestras. Experienced conductors, for example, tend to move completely differently from the basic beat pattern within a lively $^{3}/_{4}$ time, and change their metric gesture into a more circular movement (like in the beat pattern of a $^{2}/_{4}$ time, but with a 2 to 1 timing difference between the energetic swinging away from the body – plus a pending phase – and the relaxed quick swinging back to the body), which corresponds much more to a possible bowing motion in this time signature. Thus, the pulse-group of this fast triple meter can be emphasized more exuberantly within the flow of an elegantly swinging movement of the conducting arm.

But beyond the description of conductors' movements, this paper aims to investigate the possibility of physically supporting assistance within the teaching of bowed string instrumental performance, thus stimulating a metrical understanding and metrical sound experience of remarkable intensity. Even when learners do not yet possess the abilities and knowledge to imbue their sound with metrical emphasis (stress) and lightness (unstress) themselves at the time of the supporting assistance, the spontaneously facilitated physical and tonal realization of metrical performance as well as the clearly audible and experienceable difference to the learners' previous attempts should be evident and boost metrical understanding. The central cue to high artistic expertise in metrical performance is to do the emphasis without congealing stress (which often leads to a noticeably disturbing acceleration of the bowing speed or too much static pressure on the bow) but to swing the metrical weight into the string instrument by means of a relaxed movement enriched with only the necessary amount of energy.

For a better understanding of the possible physical assistances of a teacher, which will be finally demonstrated in detail within a concrete musical example, the following will provide broader considerations concerning metrical performance in general as well as a short summary of the necessary technical knowledge about bowed string instrumental performance. To provide this special kind of physical assistance in metrical bowing presented in this article, a teacher needs to be convinced of the advantages of a leverage system in string instrumental bowing that is based on a logical combination of biomechanical possibilities and sound physical requirements. Therefore, the theoretical approach in this article is complemented by a second article presenting the results of an empirical study on the relevance of mass balancing oscillations in the bowing of adult professional violinists as a characteristic element in the applied leverage system.

## 2. Metrical Tonal Shapes in Theory and Practice

Although the "metrical tonal weight" is considered theoretically only as an "ideal value", "which is attributable to a tone through its relative relationship to at least one further tone"[4] in musical practice, the ideal metrical positions of a tone become reality in sound, with effects on the putatively accurate parameters "tone duration", "loudness" and "timbre"[5]. Every tone has an individual course, an articulated shape, which is based on the interaction between these parameters. The course of loudness must harmonize with corresponding timbres, and each tone must possess a speaking, singing, or however gripping "groovy" articulation. Such a tonal shape extends theoretically to a clearly defined time - determined by rhythm and tempo – but it can already have passed into the "niente" of a not noted but metrically appropriate sound letup, before the next tone begins. The durations of tones and pauses, which are supposedly strictly fixed within the notation, are therefore often metrically relativized. Theory-based attempts at visualization of metrics via abstractly quantifying degrees of emphasis are a construction coming close to the weighting of keystrokes in piano playing, but are quite far from the practice of bowed string instrumental performance.

For example, originating from Baroque music, it is necessary to follow the initial stressed or unstressed energy pulse – which is necessary to evoke a string oscillation of the intended sound amplitude – with an adequately fitting, rounded dampening curve. Over the course of a piece of music, the impulses as well as the affiliated dampening curves of every tone must be put in relation to each other, according to the degree of emphasis in regard to their respective metrical position.

The metrical shaping of each tone in itself and in its relation to other tones challenges musicians of all genres and on all their respective instruments to adequately perform the required transitions from metrical weight and emphasis to unstressed lightness, and vice versa, according to the style of music. The metrical structure of a piece of music and thus the necessary metrical shaping of tones in their relationship to each other need to be "understood" first of all. Therefore, within most musical styles, the time system seems to pre-set an always consistent measure. The music, however, evolves vividly over melodious phrases through harmonious developments, forming larger units of meaning with influences on the metrical structure inside.

---

[4] Translated from: https://de.wikipedia.org/wiki/Metrum_(Musik), 2016-3-4.

[5] Timbre has been called, "...the psychoacoustician's multidimensional waste-basket category for everything that cannot be labeled pitch or loudness." (McAdams and Bregman 1979, p. 34; cf. Dixon Ward 1965, p. 55 and Tobias 1970, p. 409), https://en.wikipedia.org/wiki/Timbre, 2017-6-24.

With regard to the possible complexities of metrical structures, it can help at first to begin with the most basic metrical difference: that between emphasis and lightness. How can stressed and unstressed tones be created on the instrument? If, subsequently, the degrees of emphasis can be further differentiated within various time signatures and musical works or genres, it should finally lead learners to capture possible contents of the construct "meter" in their respective musical practice.

Every tone arises in the mind and is pre-determined in the body preparation before it sounds. Depending on how much energy and which kind of energy has been provided previously, the resulting tonal experiences differ. For example, the further the arm and especially the elbow rises at the tip of the bow before starting the up-bow, i.e., the more potential energy is stored up in the arm, the higher the speed of the bow driven by the kinetic energy of the relaxing arm, i.e., the momentum of transformed energy – and thus the generated loudness. The pressure of a pendulum is higher the closer the arm was previously to the horizontal before it is left to fall almost freely with the earth's gravity (cf. Mantel 2011, p. 169). And the more deeply relaxed the mass of the arm can convey its energy to the bowed string instrumental system – with the least possible amount of impeding muscle activity –, the more 'weighty' the mass can effect into the depth of a string, and thus may produce 'darker' timbres. Of course, the articulatory and coloring influences of intended muscular impulses superimpose to this necessary basis on a higher artistic level, but a successful transferal of energy from the body to the instrument in this basic manner is of decisive importance for further artistic development. Below, the considerations will focus even more detailed on the playing of bowed string instruments. Nonetheless the intention is to encourage readers from other specialist areas to comprehend these ideas and principles and to incite profitable transfers to other instruments and fields of experience the readers are more familiar with.

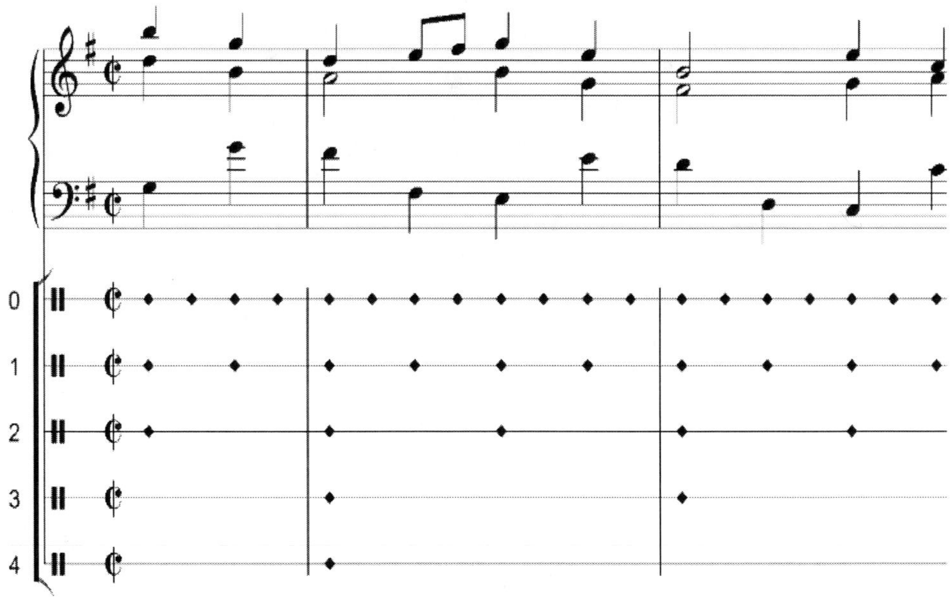


Figure 1. Attempt at visualizing metrical relations over quantifying degrees of emphasis, J.S. Bach, French Suite No. 5, BWV 816, Gavotte, time 1-2, taken from https://de.wikipedia.org/wiki/Metrum_(Musik), 2016-3-4.

After a keystroke on the piano the oscillating strings are bound to their physical dampening[6] and therefore achieve an inevitably formed tonal shape. A string tone however is theoretically still, at any point, effectible in its development by the bow and therefore unfortunately also distortable. It is a characteristic of the art of bowed string instrumental performance to be able to minimize or respectively guide this influence appropriately.

## 3. Metrical Tonal Shapes and their Movements in Bowed String Instrumental Performance

> "The musculature holding up the arm is relieved of the amount of pressure that rests on the string; the louder the sound the more relaxed this musculature will be" (Gerhard Mantel)[7].

The movements described herein refer to a leverage system of swinging mass parts in bow and arm – biomechanically organized around four pivots, according to lever laws – developed for string instrumental playing in the first half of the twentieth century by August Eichhorn[8] (cf. Hasselbach 2012, Mantel 2011, Hopfer 1941). Eichhorn's playing system could be called "counter swing technique", while its muscularly more strenuous alternative could be called "parallel leading technique", as in the course of the latter, the masses of arm and bow are moved rather up and down in parallel without clearly recognizable pivots.

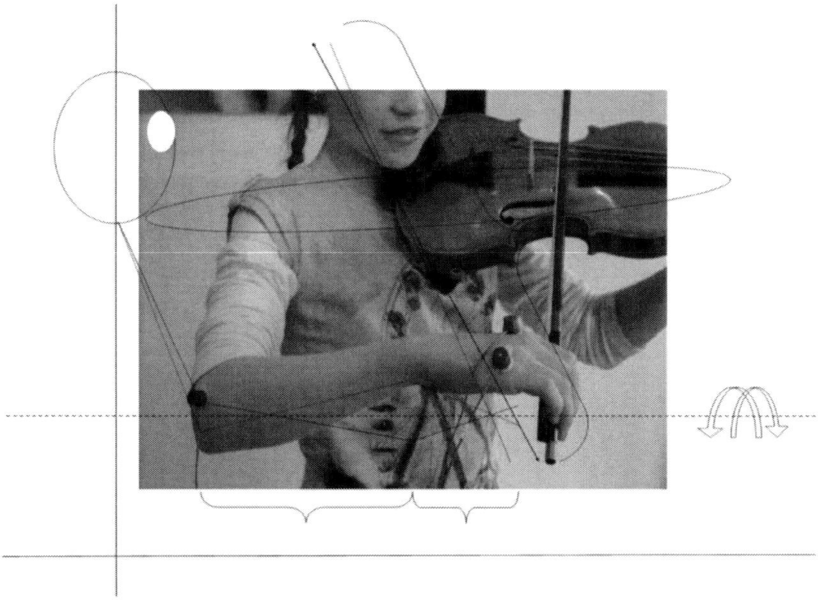

Figure 2. Schematic depiction of August Eichhorn's theory of oscillating levers in a test situation. Three ultrasound markers transfer the movement data to a computer. (Previously published in *Arts BioMechanics*, vol. 1, #1, 2011).

[6] If not additionally dampened by any kind of artistic releasing of the piano keys and thus of the dampers.
[7] Mantel 1995, p. 144, German original: Mantel 2011, p. 145.
[8] August Eichhorn (1899-1980) was a solo cellist at the Gewandhaus Orchestra Leipzig and professor of Gerhard Mantel and Josef Schwab et al., cf. https://de.wikipedia.org/wiki/August_Eichhorn_(Cellist).

Apart from the evident quasi "horizontal"[9] respectively "transversal" playing plane visible in any kind of bowed string instrumental performance, in the counter swing technique, the less visible perpendicular movement planes (quasi "sagittal" and quasi "frontal") in arm and bow are combined. In these perpendicular planes, parts of the arm of approximately identical weight oscillate in repetitive movements around a pivot and are balanced out with each other.

According to the principle of a seesaw, the mass inertia of two complementary oscillating mass parts is minimized. This movement is comparable to the everyday movement when sprinkling salt from a saltshaker onto an egg. In the counter swing technique, this perpendicular movement is superimposed on the horizontal whole bow movement in a large form while the player makes whole bow strokes. But it can also be repeated in smaller forms at any time and at any point of the bow to play faster repetitive strokes. This movement would be similar to that when sprinkling salt all the way, from an egg in front of one's own body (at the frog of the bow) to the egg on the neighbor's plate (at the tip of the bow). The direction of the horizontal movement (down-bow or up-bow) can be changed at any time while just changing the corresponding perpendicular movement's direction as well. Vice versa, the perpendicular movement can be superimposed with an additional amount (e.g., to boost the bowing speed, for string crossings or bounced articulations) or it can be reduced to a minimum at any time. Only a complete stopping of the perpendicular movement would be disturbing to the horizontal movement while using this leverage system because this would block the swinging system.

Furthermore, "dabbing" muscular energy impulses in the frontal movement plane are of great importance. These impulses are generated from muscles that actuate the forearm rotations. This energy is only transmitted to the bow by the hand's firmness and not created in the fingers. In a certain way, this movement can be called a "motor" of the swinging leverage system. The swinging of the bigger arm parts is considered to be partly reactive to these impulses. The frontal impulses are in principal independent of the transversal and sagittal movements and can be freely combined for the realization of any artistic intention. But the movements react to each other like the parts of a swinging perpetuum mobile. The better the impulses fit to the movement already stored in the system, the easier it is to keep it swinging. On the other hand, inappropriate impulses lead to inhibitions of the swinging system; accordingly, the lack of perpetuating impulses already leads to a steady loss of energy in the swinging system. The system is further complemented and kinematically closed by the instrumental movement axis of the bow, which also oscillates in the frontal plane around its pivot, the floating point of contact on the strings.

Mass percentages do not always have to be distributed evenly around a pivot. In keeping with the image of the seesaw on a playground, depending on both the distribution of masses around the pivot of a seesaw and the strength of an additional initiating impulse, the momentum and the strength of one side of the seesaw hitting the ground can be influenced. In the case of bowed string instrumental performance, the same principle works to dose the energy that is delivered to a string of the instrument, specifically in the moment of a change in the bow's horizontal movement direction, which simultaneously comprises a change of the perpendicular movement's direction. This provides the opportunity to start a tonal shape with

[9] "Horizontal" should be understood relatively and means the respective playing planes on the strings of a bowed string instrument, even though these are variously positioned in relation to an absolute horizontal.

a specific amount of energy by arranging masses, levers and pivots adequately. That is, by the way, why it is important to begin a bow stroke out of a movement simulating a previous stroke with a certain speed, a swinging direction and provident mass balancing.

In an attempt to use words that are as clear as possible in the pedagogical communication with students, August Eichhorn called the most prominently observable swinging movement of the bow arm in the sagittal plane the "Längshebel" (longitudinal lever)[10], the forearm rotation with its transversely dabbing impulses on the bow in the frontal plane the "Querhebel", (transverse lever) and the motion axis of the bow the "Bogenhebel"[11] (bow lever). The longitudinal lever, due to the great inherent mass of the arm (from the shoulder to the hand, per artistic requirements to the wrist or fingertips), usually provides the larger forces, while the transverse lever (from the elbow to the fingertips) is more concerned with precision work and force transmission. The functions and effects of the bow lever will be mostly disregarded at this point for the purpose of simplification.

Upon closer inspection, there is a type of "connecting lever" (from the wrist to the fingertips) between the transverse and longitudinal levers, a kind of "coupling". This connecting lever to the bow usually equalizes the height differences of the oscillating longitudinal lever except if it is used as an extension of the longitudinal lever, and it thus transfers the height differences directly into the bow lever, as for instance, is necessary within the bow-stroke called *Riccoché* and other bounced articulations. On the other hand, a very smooth and complete equalization of the height differences leads to bow-strokes articulated non-bounced in *Legato*. Height differences are also surmounted to change the bowed string when playing in the upper part of the bow, where the contact point of the bow on the string forms the pivot of the bow lever. Using the longitudinal lever for string crossings means to move it conjoined to the bow lever, turning both around their pivots, with a very small energy impulse without having to lift the entire arm mass.

All four levers swing interconnectedly with each other, and can only do so if all levers are movable around their pivots. If one of the levers is blocked, the others cannot swing freely enough. These levers are, from an organic viewpoint, partly identical but differ functionally due to their movement directions and their driving musculature. Artistic sound design and metrical tonal shaping in bowed string instrumental performance can be considered as a complex interaction of these levers, with constantly changing degrees of involvement of the individual levers in composited movements.

If teachers try to help students to understand meter via physical assistance and supported metrical sound experience, they have to know how these movements can be combined to form metrical tonal shapes. They have to know the leverage system in detail because a swinging movement is not a constantly guided movement. The movements must be sensibly prepared and released: Which is the best starting position of the limbs? Where are the ends of two-arm levers, on which an additional force can be supportive? In which direction and in what strength should the energy impulse be exerted? Which target position of the limbs should be achieved to be in a good position to start the next movement? It is a game of exchanging energy. Potential energy is converted into kinetic energy and vice versa. Thus, a leverage system of swinging masses needs only small energy impulses to determine the

[10] Comprising the less visible movement of the upper arm mostly in a frontal plane.

[11] According to the oral tradition of Eichhorn's former student, Prof. Josef Schwab, University of Music ‚Hans Eissler' Berlin (*1934).

movement directions. The better the movements are planned, the less inhibitory holding energy is needed to achieve the exact movement timing necessary while making music.

Theoretically well prepared, metrical tonal shaping in its basic features of emphasis (*thesis*) and lightness (*arsis*)[12] can now hopefully be comprehended in its practical realization in bowed string instrumental performance:

In the *thesis* (emphasis), the longitudinal lever, that provides arm weight and boosts the bowing speed, takes the more active part of the energy supply into the string for "dark", "deep" and "heavy" sounding timbres in combination with greater volumes. The transverse lever first helps to transfer energy well dosed into the string and then reduces the transferred energy to form a metrical tonal shape. Towards the end of the thesis – which begins heavily but ends lightly – the effective weight of the bow is even reduced by supinating[13] the transverse lever. In the transition from emphasis to lightness, "lifting" the bow allows the creation of timbres that sound "lighter".

To play a *thesis* in the down-bow from a position close to the frog, the arm's mass is lowered a little from the relaxed shoulder[14], and is immediately swung away from the body in a lateral upward movement like a pendulum. Thus, the tone begins with a saturated heaviness without "scratching" and elegantly ends in lightness. For this, the impact of the levers is combined with a convenient centrifugal force of swinging masses.

To play a *thesis* in the up-bow from a position close to the tip, stored potential energy of the whole arm is needed with the arm being in a relatively high position (the pressure of a pendulum). This potential energy is, or has been, previously accumulated in the rousing down-bow or is provided by a similar starting position. When the potential energy is released again, the arm almost freefalls towards the body, and only the hand-sided lever arm guides the bow – in a complementary movement to the elbow-sided lever arm – further upwards to balance the masses and to transform some of the kinetic energy into potential energy again.[15]

In the down-bow, the decrease of the bow's effective weight towards the tip and the rapidly swinging away of the attached arm weight are mostly sufficient to form the metrical tonal shape of a *thesis*. But in the up-bow, an equivalent is needed, one that is not only strong enough to compensate for the increase of the effective bow weight towards the frog, but is suitable as well to neutralize arm weight and to further lift the bow towards the end of the tonal shape. To accomplish this task, the transverse lever and the connecting hand lever are

---

[12] In the realm of poetry, emphasis is associated with a higher pitch; therefore, the antique terms *thesis* (lowering) and *arsis* (lifting) have been swapped. In instrumental music, emphasis is independent of pitch, so that an emphasized tone with a "stressed" metrical rank is still called *thesis*, and the "unstressed" tone *arsis*. It seems unfavourable, but hardly avoidable, that digital metronomes use a higher pitch for the *thesis* even in the context of instrumental music because the sound quality of the small loudspeakers of actual devices do not allow for clear dynamic and thus metrical differences. On smartphone apps, less than 10% of all variants use the possibility of taking a lower pitch for the *thesis*, which does not really seem to be more helpful anyway. New solutions should be developed using smartphones connected to powerful speakers that perform metrical tonal shapes, maybe even allowing for individually adjustable parameters (such as interactive graphical shaping of volume and timbre) and adaptable to different instruments and musical styles.

[13] In a supinating movement of the lower arm, with right-handedness, ulna and radius revolve around each other in a clockwise direction.

[14] The lowering and additional relaxation serves to get in increased contact with the string while proportionally increasing the effective arm weight and the bowing speed. This process of gently creating a string's amplitude is called "proportional pressure increase".

[15] Here, violin playing differs from, for example, cello playing, because with the smaller violin it is almost sufficient to let the elbow continuously sink deeper while the hand rises. However with the much bigger cello, the arm and elbow have to follow the hand further upwards, or rather on to the instrument's middle, to fully come to the frog.

"fixed" to the longitudinal lever in the sense of a reduction of the degrees of freedom in the joints. Thus the whole arm is acting as one powerful two-armed lever with the upper arm mass easily levering the masses of the lower arm and bow beyond the center of rotation. This process is also referred to as secondary supination (cf. Mantel 1999): The passive rotation of the lower arm, hand and bow through an active rotation of the upper arm in the shoulder joint – in principle nothing other than the utilization of the longitudinal lever. Hence, the functionality in the leverage system depends on the quality of interconnections between the levers, of the degrees of freedom in the joints, respectively.

In the *arsis* (lightness), the transverse lever takes the more active part in transmitting energy into the string, while the longitudinal lever supports the dosing of energy and the forming of the metrical tonal shape by lowering and lifting the bow-side of the two-armed lever. "Light" tones in a metrical sense basically have a comparable tonal shape to "heavy" tones. They just have a smaller, less voluminous corpus of heaviness and decrease more quickly. The metrical sound gap following an *arsis* is often bigger than that following a *thesis*. The lifting of the bow in an *arsis* occurs so early that the dampening curve leads to barely noticeable amplitudes of the string oscillation, often perceived as a sound pause.

To create an *arsis* a "dabbing" movement in the transverse lever provides a pronation impulse[16] onto the bow. Due to the elasticity in the system of bow-stick, bow-hair and string, the energy of the pronation impulse is reversed as on a trampoline, and the restored energy helps to lift the bow, ideally supported by an additional supination of the longitudinal lever[17]. Thus, the bow may even be spiritedly catapulted upwards – to then fall into the heaviness of the next *thesis* even more emphatically.

Below the center of gravity of the bow close to the frog, the longitudinal lever seems to suffice to perform tender "dabs" with its pronation and supination out of the whole arm rotation in the shoulder joint, which is equal to a secondary rotation of the transverse lever. But the transverse lever still gives slight impulses, even if it apparently loses its dominant role there because of an exchange of biomechanical functions[18]. In any case, it is observable that the better the masses are balanced out around the pivot of the longitudinal lever, the smaller the necessary muscular impulses are which cause the lowering and lifting of the bow according to the seesaw principle.

In summary, the movements creating metrically shaped tones usually are, on the one hand, big and easily visible swinging movements of the whole arm, be it away from the body or back again (*thesis*), and on the other hand, smaller, less visible miniatures of this movement (*arsis*), at whatever point or direction of the bow they are required. The

---

[16] In a pronation impulse of the lower arm, with right-handedness, ulna and radius are first turned counter clockwise for delivering a vital energy impulse, before swinging back clockwise to their muscularly relaxed physiological position.

[17] In the down-bow, the supination of the transverse lever partly works against the automatically given secondary pronation of the outwardly swinging longitudinal lever. However, in the up-bow, the primary supination is supported by the secondary supination in the swinging back of the longitudinal lever. This is why a quick succession of bound *Staccato*-strokes in the down-bow is more complex than in the up-bow. In the up-bow, primary and secondary supinations are added in the same direction, so that the rapid succession of pressure reliefs between the pronation impulses is more easily achieved. For the same reason, the preferred realization of metrical shaping in a minuet is with a down-bow followed by two up-bows.

[18] Below the center of gravity of the bow close to the frog, the transverse lever obviously serves more actively to rotate the bow across the strings for string changes, whereas above the center of gravity of the bow up to the tip, the longitudinal lever obviously serves more actively to rotate the bow across the strings. In any case, all levers act together, but because of the diverse lengths of lever-arms, they seem to be more or less crucial and causal to the respective torques and their energy supply.

"miniature" can, in principle, take on every possible size and even the size of a whole bow stroke if necessary, in the sense of a special case. Whether or not this produces a *thesis* or *arsis* is decided through the quantity and duration of the influence of arm mass, which means energy exertion into the "depth" of a string, creating larger amplitudes.

Basically, it relies upon the contact point of the bow on the string, which amount of effective weight and elasticity of the bow must be taken into account. The bow provides constantly changing physical conditions, and also the positions of arm limbs and joint angles in relation to the instrument are constantly changing biomechanical conditions. Therefore, it seems obvious that the more at the tip of the bow an *arsis* occurs, the bigger the amount of additional muscular energy needs to be. But using a leverage system in which masses are perfectly balanced out against each other equalizes such differences for the most part. The system provides matched preloaded energy at any time, so that the necessary additional muscular impulses are always small and nearly equal. This allows for a very reliable artistic performance. Even if the music demands varying strengths of impulses, related to different formations of amplitudes, and varying dampening curves for the desired musical expressions, the leverage system gives orientation in a reliable reference system of force relations. Using the leverage system opens up the comfortable realization of all possibilities in a metrical continuum between emphasis and lightness, independent of the bowing direction and the contact point of the bow, with volumes and timbres melting together into characteristic tonal shapes within the flow of movement.

## 4. Physical Assistance and Supported Metrical Sound Experience as a Method in String Instrumental Pedagogy

Physical assistance – be it purely non-verbal or accompanied by explanatory words – can be an important methodological step towards metrical sound experience and further processes of understanding aspects of musical meter. This step might be useful both at the beginning of a learning process and as a recurring stimulus. Especially in interplay with an accompanying theoretical approach and with the perception of a teacher's model performance, the supported experience of a semi self-done high-grade performance can be conducive for learning. The body of the student as an organ of perception, reflection and cognition is thereby used in the sense of a "silent learning" from body to body, solely accompanied by metrically meaningful musical tonal shapes.

An indispensable requirement for the possibility of physical assistance in bowed string instrumental performance is the consensus on the existence of "ideal movements" in terms of a technical guiding principle. The more concrete and scientifically founded the guiding principle, the better warranted is the outcome of a physical assistance. Eichhorn's biomechanically and sound-physically substantiated leverage system seems to be preeminently suitable. By means of well-dosed physical assistance, learners should come closer and closer to the experienced ideal movements within their autonomous attempts. It is characteristic for this kind of learning processes that at the beginning there is a vast influence of the teacher on the central parameters, such as starting position of limbs, joint angles, swinging directions, torque energy, bowing speed, contact point, and pressure of the bow on the string over the whole course of the metrical tonal shape. This influence is reduced with the progressive takeover by appropriate activities of the student.

The performance under physical assistance should finally induce a successful autonomous performance. Looking at this from a motor learning viewpoint, the method intervenes in motor learning in its control-loop of “movement objective, movement and motion perception” (Mantel 2011, p. 14; cf. Figure 3) as well as in its control-loop of “musical objective, movement objective, movement, tone, listening” (ibid.). On the one hand, during a performance under physical assistance, the movements to be learnt can be felt inside the own body – not only via mirror neurons – and the movement perceptions can be directly compared and saved in the body and brain. On the other hand, by listening to the associated sound results in the own instrument, the students understand better the formed metrical tonal shapes as musical objectives. By changing the movement via physical assistance, the teacher can manipulate the central element in both control-loops with effects of accommodation in the challenged schemata of objectives.

The motor learning can, in addition, be profitably supported through the cognitive acquisition of knowledge about relevant biomechanical and sound-physical aspects of bowed string instrumental performance. But detailed knowledge about Eichhorn’s theory on the part of the student is not absolutely mandatory. The pure method of physical assistance is therefore especially advisable and appropriate for children up to the age of eleven, whose cognitive maturation does not allow for a deeper abstract thinking and thus the understanding of physical and physiological systems. However, on the part of the teacher, substantial knowledge about the theoretical background, in particular Eichhorn’s leverage system with its related artistic aspects, is an essential prerequisite for successful physical assistance: Without knowing where and how to give helpful impulses for setting the leverage system of arm and bow into swinging actions, the desired sound results will not occur, and the intended metrical sound experiences will not be achieved.

At first this kind of physical assistance demands almost completely passive movements on the part of the student. Little by little through stimulating movement images, such as “immersing deep into warm water” and “swinging like a pendulum,” or through more concrete explanations and descriptions concerning the correlations between movements and artistic outcomes, the initial passivity comes closer and closer to an active reproduction and finally an artistically autonomous performance.

In the first more passive stadium of learning, paramount attention is paid to raising the awareness of existing unfavourable motor imagery and relaxing the usually manifold muscular tensenesses that inhibit free swinging and the use of gravity (cf. Prelle 2015). But beyond that, it is finally a matter of developing a pronounced feeling for the right directions and speeds of swing, for the measures of mass insert and of lever forces, in order to obtain the desired metrical tonal shapes in adequate sound quality.

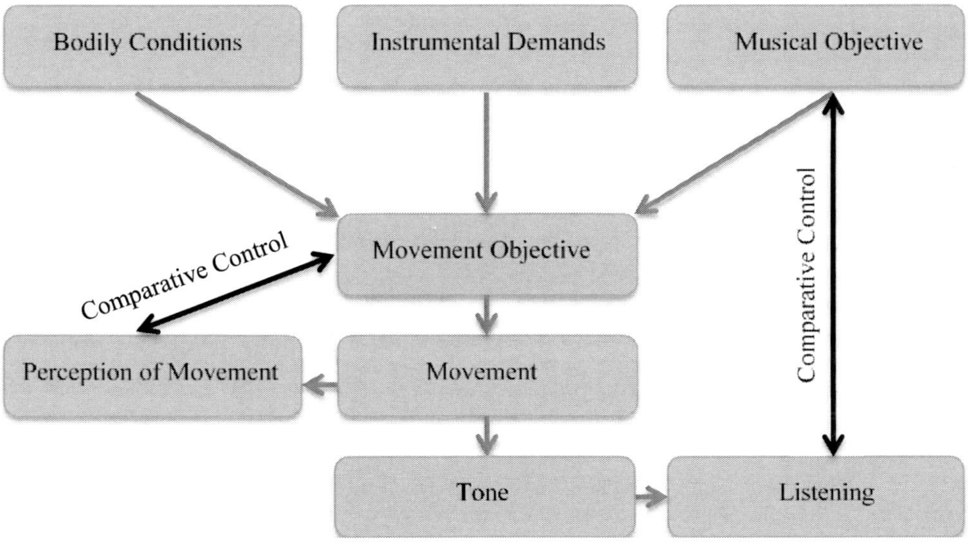


Figure 3. Organization of musical motor control according to Gerhard Mantel, taken from: Mantel 2011, p. 14, terms translated by J.v. Hasselbach.

## 5. Physical Assistance by Example of Violin Bowing in Triple Meters

When an amateur musician is asked to play the typical pulse-group of a triple meter[19], or metrical tonal shapes in a sequence of *thesis-arsis-arsis* (emphasis-lightness-lightness), the resulting sound is mostly not one of "bulbous" depth at the beginning of the *thesis* and an "organic" fading away. The rather equally "flat" tones are marked much more by a stopping in the flow of movement and are thus separated from each other. The duration of tones contributes more to a differentiation than differences in volume and timbre. It is particularly unfavorable for a tonal shape when the bow does not leave the string towards the end of a tone. Just stopping the bowing speed harshly stops the string oscillation and "chokes" the sound. At this point of missing expertise, physical assistance can be offered. The following is a detailed description of the biomechanical processes needed to form pronounced metrical tonal shapes. This exemplary template can hopefully be comprehended and tried out by experienced teachers with sufficient imagination.

In a preparatory phase, the positions of the involved limbs of the bowing arm are optimally adjusted. The working tensions of all involved muscles, especially those of the shoulder joint, are lowered to a minimum, to allow for a maximum capability of free swinging. All levers of the bowing arm – as well as the bow lever itself – are brought into a basic setting, which is advantageous for the starting of a sonorous string oscillation via proportional pressure increase (cf. footnote 15) and capable of the onset swinging from a starting position at the frog in down-bow:

[19] The most common time signatures in triple meters are 3/4 time, 3/2 time, 3/8 time or 9/8 time as a compound triple meter. Dances like the minuet, waltz and mazurka, for example, are composed in triple meter.

1. Relaxed suspension of the arm from the shoulder;
2. Deep, relaxed position of the elbow;
3. Flexible, rounded wrist turned upwards;
4. Flexible, rounded fingers in good contact with the bow (without lifting the bow with pressed and stiffened finger joints), so that the bow's weight mostly rests on the string, and hardly any muscle strength is used to lift the bow;
5. Relatively relaxed muscles of finger joints and wrist, but in a slight working tension suitable to adjust the bow at a right angle to the string (pushing or pulling the bow lever in the horizontal plane, turning it around its pivot on the string).
6. Preparatory, small swinging movements of the longitudinal lever in perpendicular direction to the string can already be tried to ensure that not too many muscles still reveal inhibitory tenseness in the shoulder joint. Ideally there is a completely relaxed shoulder, but enough working tension in wrist and fingers to extend the longitudinal lever to a whole arm lever with energy transmission directly onto the bow.

Then follows the onset of swinging in the levers through suitable energy impulses. To initiate a *thesis* in the down-bow these are the physical assistances to be given:

1. At the elbow: Pull the upper arm slightly downwards to leave the arm-weight to gravity and then swing the elbow outwardly away from the body, thereby subsequently opening the elbow-hinge-joint. The masses of forearm and bow continue the swinging movement "flying" further outwards.
2. At the bow: While transitioning from emphasis towards lightness within this tonal shape, the transverse lever is needed to complete the lifting of the bow. Support this supinating movement of the forearm by turning up the bow a little bit, grasping the bow close to the student's hand near the pivot of the transverse lever.

To then initiate the first *arsis* in the up-bow in the upper half of the bow:

1. At the bow: Give a slight pronation-impulse with the transverse lever on the bow (or just push directly) to dab energy into the string and support the bow's elastic coming back out of the string through a slight renewed turning up of the bow (primary supination of the transverse lever).
2. At the elbow: Lower the elbow a little bit and let the longitudinal lever drop into relaxation in the shoulder joint, thus lifting the hand-sided lever arm. This supports the primary supination of the transverse lever through secondary supination out of the longitudinal lever.

To initiate the second *arsis* in the down-bow in the upper half of the bow:

1. At the bow: Turn down the bow onto the string and give a slight pronation-impulse with the transverse lever on the bow to dab energy into the string and support the bow's elastic coming back out of the string through a slight renewed turning up of the bow with primary supination of the transverse lever only, while the arm continues to swing outwardly.

2. At the elbow: While pronating, lift the elbow a little bit to support the transverse lever, then let the longitudinal lever swing further up- and outwards (down-bow-*thesis*-miniature).

To initiate a *thesis* in the up-bow at the tip of the bow:

1. At the elbow: Give a quite strong impulse downwards and let the elbow swing back to the body with the help of gravity, transforming the previously stored potential energy back into kinetic energy. If the postural muscles of the shoulder joint are not ready for a quick relaxation, hold the elbow in the high position until the student's shoulder has relaxed and then suddenly let the arm mass go for a swing back in "free fall".
2. At the bow: Support the transition from emphasis into lightness by turning up the bow, using both supination of the transverse lever and especially the effect of the secondary supination out of the longitudinal lever. The transmission of secondary supination into the bow lever depends on a certain firmness of the rounded wrist, stabilizing the connecting lever, and a "clawing" the bow with the fingers in the transverse lever. This can be supported by embracing the student's fingers with the whole hand to keep them round while lifting the forearm and turning up the bow.
3. At the elbow, but with effect on the wrist (necessary, if the wrist does not get rounded by itself towards the frog): Support the flexible rounding upwards of the wrist with a little "kick" (with the thumb, while the fingers continue leading the elbow downwards) at the side of the cubit (ulna) upwards to the wrist, to better adjust the joint angles in the connecting lever and the transverse lever to remain in a right angle of the bow to the string.

To initiate the first *arsis* in the down-bow in the lower half of the bow:

1. At the elbow and the bow simultaneously: Reverse the swinging directions of the lever-arms of the longitudinal lever through simultaneous impulses at the complementary points (elbow and hand/bow). Use the longitudinal lever to "dab" a slight energy impulse into the string via secondary pronation and let the arm further swing a little bit away from the body (down-bow-*thesis*-miniature). Support the transition from mini-emphasis into extreme lightness through secondary supination out of the longitudinal lever.

To initiate the second *arsis* in the up-bow in the lower half of the bow:

1. At the elbow and the bow simultaneously: Reverse the swinging directions again to "dab" a slight energy impulse into the string via secondary pronation out of the longitudinal lever and then immediately reverse directions again with the elastic comeback of the bow, without disturbing the flow of the up-bow movement. Let the elbow progress in swinging further down (up-bow-*thesis*-miniature), lifting the hand and turning up the bow on the other side of the lever, taking advantage of the same swinging directions of the longitudinal lever in the fundamental movement of the up-bow and its supination for secondary supination of the transverse lever.

For the realization of the two basic metrical tonal shapes of *thesis* and *arsis*, with a second *arsis* following in triple meters, six different movement-combinations are necessary in this example – dependent on the bodily conditions and the string-instrumental demands – before the pulse-group starts again with a *thesis* in the down-bow. This performance with a permanent alternating in the *thesis'* bowing direction is one of the most common bowings in triple meters. But the bowing used possibly even more is to start every pulse-group with a down-bow for the *thesis* and always playing the second *arsis* in the lower half of the bow in a second up-bow. Thus, after the first two metrical tonal shapes in the example above (a *thesis* in down-bow and an *arsis* in up-bow in the upper half of the bow), the last movement of an *arsis* in the up-bow in the lower half of the bow is added. This bowing gets along with only three different movement- combinations, so that every pulse-group is bowed identically.

## Conclusion

If there is sufficient awareness of possible fundamental combinations of movements for a metrically differentiated string instrumental performance, sequences that save energy and reduce complexity in bowing can be chosen. The acceptable freedom in musical interpretation normally allows for a reasonable arrangement of the bowing in a musical work. Using the benefits of a leverage system, like the one Eichhorn invented for string instrumental performance, may have a huge influence on the success of the artistic performance as well as on the bodily exertion and stress that can come about through professional string instrumental performance.

Whichever "ideal movements" in terms of a technical guiding principle dominated a teacher's learning biography in string instrumental performance and characterize his or her teaching at present, the body with its perceivable movements and the experienced sounds associated with the movement can be understood as indispensable parameters in an hermeneutic process of understanding music, exemplified here in the field of musical meter. Physical assistance can intensify and possibly accelerate the learning process, as the given theoretical background allows for the definition of "ideal movements", which have been partly investigated in empirical studies to verify their relevance (cf. Hasselbach et al. 2010, 2011, 2018). Yet the cognitive-intellectual foundation of understanding musical meter within lessons of music theory and an emotional anchoring in music making, remain indispensable driving forces of performances. In a consciously reflected interaction of gravity, lever forces, torques, muscular energy impulses for swinging masses and energy transformation with space, time and sound, the understanding of musical meter becomes manifest within the flow of movement and can be experienced simultaneously both in music and in the body.

## References

Hasselbach, J. v., Gruhn, W. & Gollhofer, A. (2018). Mass balancing oscillations in the bowing of adult professional orchestra violinists: prevalence, tempo-related profiles and their relation to occupational health, *Arts BioMechanics, Volume 2* (number 2), the previous article in this issue.

Hasselbach, J. v. (2017). Schwere und Leichtigkeit im Fluss der Bewegung: Metrisches Verstehen und Klangerleben im Streichinstrumentalspiel. In L. Oberhaus & Chr. Stange (eds.), *Musik und Körper: Interdisziplinäre Dialoge zum körperlichen Erleben und Verstehen von Musik* [Gravity and lightness in the flow of movement: Metric understanding and sound in string instrumental playing. In L. Oberhaus & Chr. Stange (eds.), *Music and the Body: Interdisciplinary Dialogues on the Physical Experience and Understanding of Music*] (pp. 121-143). Bielefeld, Germany: [transcript].

Hasselbach, J. v. (2012). For 'lively' and 'intensely related' tones in bowed string instrumental performance: A response to Peter Röbke's introduction to "Das Musizieren und die Gefühle," *Arts BioMechanics, volume 1* (number 2), 115-130.

Hasselbach, J. v., Gruhn, W. & Gollhofer, A. (2011). Effects of training on mass balancing oscillations in the bowing of (pre)teen violin students: A quantitative micromotion study, *Arts BioMechanics, volume 1* (number 1), 1-14.

Hasselbach, J. v., Gruhn, W. & Gollhofer, A. (2010). Mass balancing oscillations: An indication of expertise in the bowing of violinists: A quantitative micromotion study, In S. M. Demorest, S. J. Morrison, P. S. Campbell (eds.), *Proceedings of the 11th International Conference on Music Perception and Cognition (ICMPC 11,* pp. 388-393). Seattle, USA: University of Washington.

Hopfer, M. (1941*). Die Klanggestaltung auf Streichinstrumenten: Das Naturgesetz der Tonansprache: kurze Einführung in "Die gestaltende Dynamik der Bogenmechanik" von August Eichhorn* [*The Sound Design on String Instruments: The Natural Law of the Sound Speech: Brief Introduction to "The Forming Dynamics of Bow Mechanics" by August Eichhorn*]. Leipzig, Germany: Kistner und Siegel.

Mantel, G. (1995). *Cello Technique: Principles and Forms of Movement* (translated by Barbara Haimberger Thiem). Bloomington, USA: Indiana University Press.

Mantel, G. (2011). *Cellotechnik: Bewegungsprinzipien und Bewegungsformen* [*Cell technology: movement principles and forms of movement*] (revised edition). Mainz, Germany: Schott.

Mantel, G. (1999). *Cello üben: Eine Methodik des Übens nicht nur für Streicher* [*Practicing Cello: A technique of practicing not just for strings*] (supplemented edition). Mainz, Germany: Schott.

Mozart, L. A. (1985). *Treatise on the Fundamental Principles of Violin Playing*. New York, USA: Oxford University Press.

Mozart, L. (2007). *Gründliche Violinschule* [*Thorough violin school*] (First edition in modern writing and adapted spelling of the second edition of 1769). M. M. Beckmann (ed.). Salzburg, Austria: Kulturverlag POLZER.

Prelle, U. (2015). *Leichtigkeit: Eine ergänzende Streichermethodik zur Befreiung der rechten und linken Hand* [*Lightness: A complementary string method for the liberation of the right and left hands*]. Mainz, Germany: Schott.

In: Advances in Arts Biomechanics
Editors: Gongbing Shan and Peter Visentin
ISBN: 978-1-53615-641-6


*Chapter 6*

# MASS BALANCING OSCILLATIONS IN THE BOWING OF ADULT PROFESSIONAL ORCHESTRA VIOLINISTS: PREVALENCE, TEMPO-RELATED PROFILES AND THEIR RELATION TO OCCUPATIONAL HEALTH

***Julia von Hasselbach[1], Wilfried Gruhn[2] and Albert Gollhofer[1]***
[1]University of Freiburg, Germany
[2]University of Music Freiburg, Germany

## ABSTRACT

Mass balancing oscillations (MBOs)[1] have already been investigated as an indication of expertise in the bowing of violin students at music universities, but not yet in the bowing of professional orchestra violinists who, beyond questions of expertise, have to cope with high rates of playing-related musculoskeletal disorders (PRMDs). The extent of using MBOs may be a considerable biomechanical risk-reducing factor.

This study intended, first, to gather data on MBOs from adult professional violinists working in German orchestras and, second, to investigate the relation between MBOs and PRMDs. Therefore, data on MBOs in repetitive bowing movements of orchestra violinists were captured three-dimensionally and quantified by computer analysis at various tempi. The resulting individual tempo-related profiles of MBOs were analysed qualitatively according to Eichhorn's theory of a leverage system in string instrumental bowing. Recorded structured interviews were held to obtain reliable biographical data.

The measurements of 19 participants (aged 23 to 62 years) resulted in a mean of 61.3% mass balancing oscillations at high tempi (MBOs/HT). The analysis of differences in the individual tempo-related profiles of MBOs suggests five types that were repeatedly found. There was a significant negative correlation between MBOs/HT and the "suffering from a PRMD" as well as between MBOs/HT and the sensation of "subjective impairments in violin playing due to a PRMD." A one-way ANOVA confirmed the dichotomous factor "suffering from a PRMD" to be decisive for a significant mean difference of 13.6% MBOs/HT between healthy violinists and PRMD affected violinists.

---

[1] MBOs have been previously referred to as CM (complementary motion) in prior publications. This abbreviation is changed for a better understanding.

Pronounced mass balancing oscillations in string instrumental bowing especially at fast tempi may be of great importance both to the artistic development and to the musculoskeletal health of professional orchestra violinists.

## Background and Aims

Bowing technique is of much importance for an artistic development that is necessary for high professional requirements. But which "art of bowing" is the best? Violin teachers tried to find supporters of their personal suggestions for bowing movements already in the 19th century, using words like "secret" or "unfailing" to praise their descriptions (cf. Sass & Stoeving 2017/1907). Noticeably teachers' opinions at the beginning 20th century were concurrent with scientific findings on physiological and biomechanical aspects of string instrumental bowing (cf. Steinhausen 1904/1928, Trendelenburg 1925/1974, Becker & Rynar 1929/1971, Hopfer 1941).

This chapter refers to August Eichhorn (1899-1980) who developed his theory of a leverage system in string instrumental bowing through the biomechanical analysis of Emanuel Feuermann's cello bowing technique[2]. "Feuermann …had the technique of a virtuoso who never had to struggle" (Morreau 2002, p. xi). Eichhorn tried to unite scientific findings with artistic phenomena and requirements and came to suggest two principles of fundamentally different movement qualities: "parallel leading technique" which is characterized by a steady muscular control of masses in bowing movements with static muscle tensions working against gravity, and "counter swing technique" (cf. Hasselbach 2018) which is comprised of "mass balancing oscillations". Eichhorn's leverage system is an example for the application of maximum mass balancing in high-performance activities. It has been described as an interconnected chain of swinging two-armed levers which allows for the greatest possible economy concerning muscular energy in string instrumental bowing. The leverage system supports an extremely precise and reliable artistic performance as is needed in the interaction with other musicians in an orchestra or as a soloist.

Since 2008 empirical studies have been undertaken concerning mass balancing oscillations in bowed string instrumental performance. A first pilot study examined six violinists studying at the Universität der Künste (University of the Arts) Berlin, Germany. In this pilot study, mass balancing oscillations were revealed as a valid indicator of expertise. The quantification of MBOs expressed the proportion of complementary swinging movements as percentage of the measurement points (cf. Hasselbach et al. 2010). The highest level of expertise detected in 2008 showed bowing techniques with 82% MBOs (among performance students); this rate dropped to 65% at the middle level (among students of instrumental education) and to 27% at the lowest level (among music education students).

---

[2] Feuermann was already teaching as a professor at Cologne University of Music at the age of 16! Eichhorn, who studied physics and physiology as well, was one of his first cello students. As Feuermann had done a lot of his vast early practicing alone, playfully copying his elder brother's violin studies and competing with him, Feuermann's technique was a phenomenon at his time: "I place him as the most important figure for 20th-century cello playing because he was the one who proceeded to carry cello playing as such to a higher level." (János Starker 1994), "The subtlety of phrasing which is so unobtrusive and so uniquely shapes the music of which he was very, very conscious in every detail. Yet when you hear a performance everything falls into place…" (Georges Szell 1967), all citations from: Morreau 2002, p. x.

This lowest level was of significant difference to the highest level of expertise, which was formed by the performance students.

Assuming a high use of MBOs to be desirable in the context of a more comprehensive expertise paradigm that focuses on the quality of movement in addition to the prevailing factor of accumulated practicing hours, it was of great interest to examine children at a young age concerning their amount of MBOs and whether or not this level could be influenced by special training based on Eichhorn's leverage system. Therefore, a second study examined 10 children and teenagers between the ages of 10 and 15. They used an average of 31% MBOs at the beginning and 56% MBOs after 15 weeks of intense special training during their violin lessons. A control group, matched in age and years of instruction, used 23% MBOs at the beginning and only 28% MBOs after 15 weeks without special training, just due to maturation and normal violin lessons. This was a significant increase in the extent of using MBOs due to the special training. Looking at the achieved peak values of up to 75% MBOs, several children were even able to reach results that were already plainly in the region of highest professional expertise (cf. Hasselbach et al. 2011). This study suggested that the artistic development of children and especially young adults can be supported by a string instrumental education that imparts biomechanical knowledge on mass balancing oscillations and on artistic sound shaping with the help of a counter swing technique.

In order to enhance the validity of the entire project, it seemed advisable to record real professional violin performances. The current article is the first that delivers data on MBOs in the bowing of adult professional violinists working in German orchestras. The results of the previous studies suggested a tentative rough classification into three segments of expertise: 20-40% MBOs = low level, 40-60% MBOs = medium level, 60-80% MBOs = high level of expertise. Therefore, it seemed conducive to investigate whether or not the data on MBOs among professional orchestra violinists lie in the assumed high level of expertise between 60% and 80% MBOs.

Finally, it was of great interest to investigate if there is a detectable relation between the extent of mass balancing oscillations at high tempi (MBOs/HT) and the suffering from playing-related musculoskeletal disorders (PRMDs) because fast repetitive movements can lead, for example, to overuse syndromes (cf. Shan et al. 2004). The susceptibility of string instrumentalists to playing-related health impairments has been proven to be very high since the 1980s (cf. e.g., Fry 1986, Caldron et al. 1986, Fishbein et al. 1988). Recent studies have confirmed the extreme high prevalence among orchestra violinists all over the world (cf., e.g., Kok et al. 2016, Berque et al. 2016, Sousa et al. 2016, Kenny et al. 2016, Steinmetz et al. 2015, Leaver et al. 2011, Kaufmann-Cohen & Ratzon 2011). As De Souza Moraes & Antunes (2012) point out in their systematic review of studies on musculoskeletal disorders among violinists and violists, among manifold other environmental and individual factors, "[t]he synthesis of the bibliographical review usually determines that the factor responsible for the musculoskeletal disorder in musicians is biomechanical" (De Souza Moraes & Antunes 2012, p. 44).

Recalling that Eichhorn differentiated between a static muscular load which predominates in a "parallel leading technique," as compared to the dynamic muscular load in a "counter swing technique" that uses slight impulses of antagonist muscles, it seemed plausible to suspect an inversely proportional relationship between the extent of MBOs/HT used in violin bowing and the musculoskeletal health of a violinist. In fast repetitive movements, force pulses have to be compensated for in their reactive direction. If force pulses are not balanced out in a swinging system of masses, abrupt accelerations in each change of the bowing direction lead to considerable vibrations of the trunk, which is countered by additional static maintenance work of the trunk muscles. Thus, high percentages of MBOs/HT may be a risk-reducing factor, both out of theoretical considerations and on the basis of numerous reports of string instrumentalists who recovered from playing-related musculoskeletal disorders by a change of the bowing movements: namely, by moving beyond a muscularly exhausting struggle to the use of mass balancing oscillations as was proposed in Eichhorn's leverage system.

## Methods

Coming from four different orchestras, 19 professional orchestra violinists gave informed consent to participate in the study[3]. Of these participants, nine violinists work in "normal" municipal orchestras, which includes playing in the orchestra pit for operas, and 10 violinists work in internationally renowned orchestras, playing symphonic music all around the world. In terms of their positions in their respective orchestras, there were six leaders (3f/3m), four principals (4m) and nine tutti violinists (4f/5m). Thus 10 violinists held leading positions and nine in tutti positions. It was difficult to obtain enough volunteers, so a random selection of participants from the orchestras, especially concerning the ages of subjects, was far out of reach.

Thanks to a mobile device, the measurements took place in a relaxed atmosphere without time pressure. The bowing movements were captured three-dimensionally with a Zebris CMS 20 measurement system[4]. Three ultra-sound-markers transmitted the movement data of hand and elbow – the two exterior areas of the possibly complementary swinging lever-arms in the bowing arm – for a quantification of mass balancing oscillations via computer analysis.

The participants had to play a G-major scale (two octaves) with four repetitions of each note. The sound was recorded in order to focus the participant's attention on a good sound quality. The scale was played at different tempi: at 60 bpm, at 120 bpm, and three times at a freely chosen higher tempo. To determine the whole group's mean value of MBOs at high tempi (MBOs/HT), the average value of MBOs for each participant was taken into account that emerged from the three trials in freely chosen higher tempi. The measurement system, the musical task and the quantification algorithm were the same as in the previous studies with students and children. This ensured the comparability of the reported percentages of MBOs.

To control for the variable "handedness," a lateralization coefficient was determined using the Edinburgh Handedness Inventory (Oldfield 1971). This preference-based inventory delivers metrically scaled percentage values that can be correlated to other variables. More elaborate tapping tests were not used although they are regarded to be more profound multidimensional hand power tests (cf. Kopiez et al. 2010, 2012, Tapping 2008).

A structured interview with each orchestra violinist was conducted individually and recorded to provide reliable data concerning PRMDs. The variable "suffering from a PRMD" was determined as a nominally scaled dichotomous variable. The variable was affirmed under

[3] I would like to thank all those involved for their support! Without the consent of the respective management teams and the participants, this study would not have been possible. It was a pleasure to meet all these wonderful people!

[4] Cf. zebris Medical GmbH, Isny, Germany, https://www.zebris.de/. Thanks also go to zebris for their support!

the condition that a musculoskeletal disorder, considered as related to violin playing by the participant, led to pain of remarkable intensity for a longer period of time, thus excluding pains of short duration and single occurrence. Skin diseases, such as inflamed pimples at the contact areas with the instrument, were not taken into account although they can be triggered and aggravated by static muscular tenseness or compensating muscular activities in the temporomandibular region. Additionally, the interview included questions that captured self-reporting statements concerning participants' satisfaction with their personal careers, violin playing techniques, physical health, and occupational stress. Standardized answers were possible in a 6-point numeric rating scale (NRS-6) to provide data tolerable as interval-scaled in these variables. Furthermore, the personal sensations of subjective and objective impairments due to a PRMD were questioned in a NRS-6 as well. A subjective impairment could be, for example, the psychological stress while trying to fulfil the artistic requirements in the orchestra despite a PRMD. An objective impairment could be, for example, the extent of sick leave from the orchestra.

The statistical methods consisted mainly of descriptive statistics and correlation analysis. The dichotomous grouping variables "type of orchestra," "position in the orchestra," "gender" and "suffering from a PRMD" were examined in detail concerning mean differences in dependant variables between the groups.

## RESULTS

The 19 participating adult professional orchestra violinists analysed as a whole group (7 female/ 12 male) were aged 23 to 62 years (mean 44.4, SD 11.6). They passed the entrance audition to their first professional orchestras between one year and 32 years ago (length of service: mean 18.8, SD 10.0 years). The freely chosen higher tempi in the musical task of this study ranged from 143 bpm to 188 bpm (mean 174, SD 10.4). Within the task of fast repetitive strokes while playing sixteenths in these high tempi, this group of adult professional orchestra violinists showed a mean of 61.3% MBOs/HT (min. 43.1%, max. 77.5%, SD 11.4%).

There was no significant correlation between the percentages of MBOs/HT and "handedness" ($r = .162$), the "freely chosen higher tempi" ($r = .163$), the "age" ($r = -.144$), the "length of service" ($r = -.217$), the "gender" ($r_{pb} = -.026$), the "instrument playing time" ($r = .040$), the "type of orchestra" ($r_{pb} = -.033$) or the "position in the orchestra" ($r_{pb} = -.296$). Nor was there a correlation between the percentages of MBOs/HT and the occupational stress level, general satisfaction with their careers, playing techniques, or health.

There was, however, a significant negative point-biserial correlation (at the .05 level) between the percentages of MBOs/HT and the "suffering from a PRMD" (Pearson, $r_{pb} = -.542$, $p = .016$). The determination coefficient $r^2$ enlightens 29.4% of the variance within the subjects of the study and estimated 24% ($\omega^2$) within the population of professional orchestra violinists. There was also a significant negative rank correlation (at the .01 level) between the percentages of MBOs/HT and the personal sensations of "subjective impairments due to a PRMD", taking also into account the subjects with no PRMD as rank zero (Kendall-Tau-b = -.578, $p = .001$) (see Figure 1). Excluding the subjects with no PRMD led to a significant negative rank correlation as well (Kendall-Tau-b = -.499, $p = .018$). But there was no significant correlation between the percentages of MBOs/HT and the sensations of "objective impairments in the orchestra due to a PRMD."

The occupational stress level at the orchestra service showed a marginally significant negative correlation with the general satisfaction with health (r = -.453, p = .052) but not at all with the suffering from PRMDs or subjective or objective impairments while playing in the orchestra. Instrument playing times showed only a slight association with the suffering from PRMDs in this study (r = .348, p = .145).

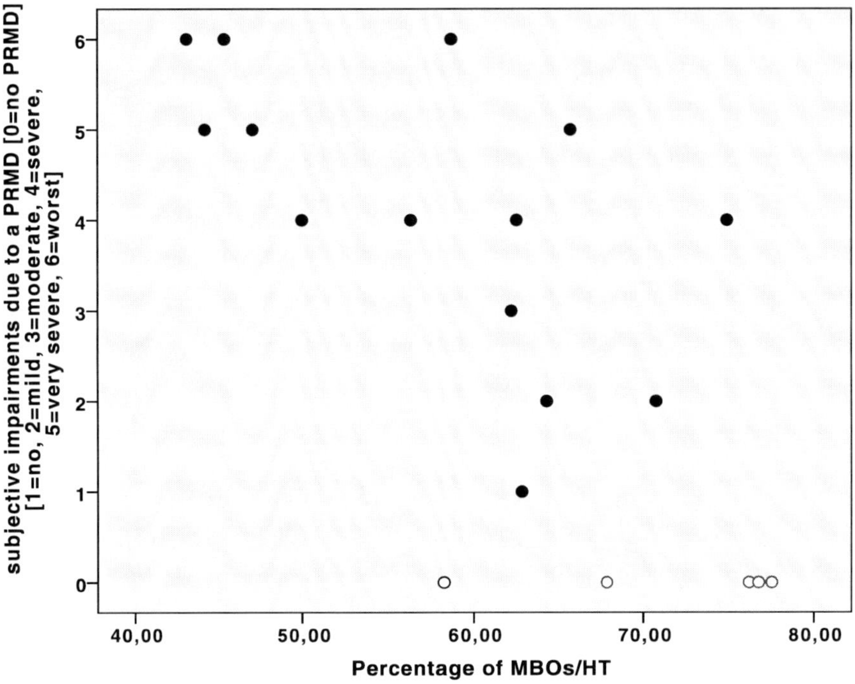


Figure 1. Scatterplot of the percentage of mass balancing oscillations at high tempi (MBOs/HT) of 19 adult professional orchestra violinists in relation to reported "subjective impairments due to a PRMD" (numeric rating scale: 1 = no; 6 = worst, 0 = no PRMD at all).

There was also a significant negative correlation (at the .05 level) between "age" and the "average freely chosen tempo" (r = -.465, p = .045) as well as between "handedness" and the "average freely chosen tempo" (r = -.527, p = .02). Violinists with less pronounced right-hand drift and younger violinists tended to play faster. There was a marginally significant correlation between "age" and the reported "handedness" (r = .588, p = .074).

Analyzing the grouping variable "type of orchestra" resulted in no significant mean differences at all between the violinists working in a municipal orchestra including playing in the pit for operas and violinists working in internationally renowned orchestras playing symphonic music all around the world. Nor did "gender" result in any significant mean differences. The "position in the orchestra" matched with participants' satisfaction concerning the used violin-playing technique (mean difference of 0.67 points in the NRS-6, p = .026). Violinists in a leading position considered their technique significantly more satisfactory than violinists in a tutti position.

"Suffering from a PRMD" resulted in a significant mean difference of 13.6% (SD 5.1%) concerning the percentages of MBOs/HT, with one-way ANOVA resulting in $F = 7.08$ ($p = .016$). The group of 14 violinists with reported PRMDs played with significantly lower percentages of MBOs/HT (mean 57.7%, SD 10.3%) than the group of 5 violinists without reported PRMDs (mean 71.3%, SD 8.3%). The Levene's test of homogeneity of variances in the groups was confirmed with $p = .433$. Control of results due to the different group sizes with a Mann-Whitney-U-Test also resulted in a significant mean difference, with an effect size of $r = .53$ ($p = .019$).

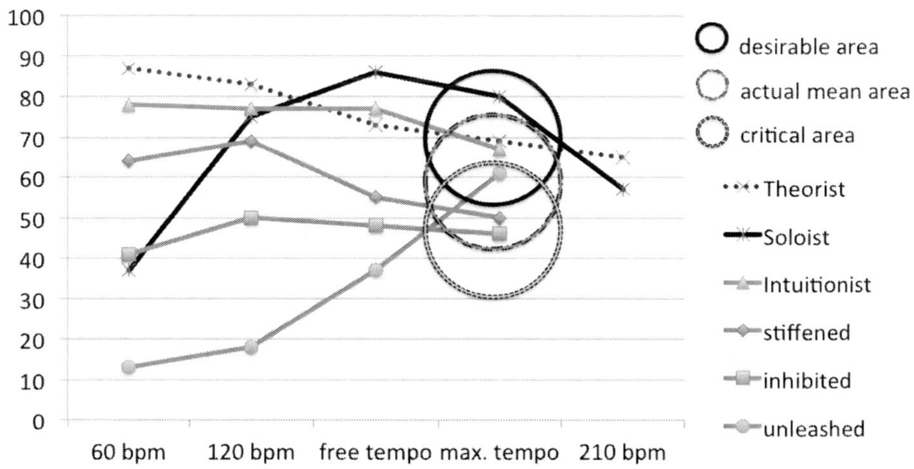


Figure 2. MBOs [%] in five categorized types of tempo-related profiles, compared to a "theorist" type that was found when a violinist played with movements according to Eichhorn's theory of a leverage system in string instrumental bowing. Every participant's tempo-related profile of MBOs in this study could be associated to one of these five types.

The qualitative analysis of distinguishing features in the individual profiles of MBOs at diverse tempi, based on Eichhorn's theory of a leverage system in string instrumental bowing, revealed recurring properties. Whereas in the slowest tempo the individual modes of motion varied widely between "parallel leading" (10-40% MBOs) and pronounced "mass balancing" (60-90% MBOs), there was an overall tendency to approach the area around the mean of 61.3% MBOs only with increasing tempi (cf. Figure 2). Depending on where the profiles started before reaching the actual observed "mean area" and taking into account the degree to which they arrived at the assumed high level of expertise between 60% and 80% MBOs/HT, the tempo-related profiles can be categorized into different types. Meaningful type names have been given to make their characteristics accessible, as is a common practice in qualitative content analysis.

Playing the musical task of this study in a test run with movements according to Eichhorns' theory of a leverage system in string instrumental bowing resulted in a profile which may be called the "theorist" type[5]. In this type, the values of MBOs drop off slowly from high perfection to some precision losses and slight stiffening at higher tempi, but still remain above 60% MBOs/HT, close to 70% MBOs/HT. A similar profile was found within

[5] Measured MBO values in playing the musical task by the author and violinist Julia von Hasselbach, who studied violin performance according to Eichhorn's theory with Eichhorn's former student Prof. Josef Schwab.

the orchestra violinists of this study and may be called the *"intuitionist"* type. The violinists in question had never heard of Eichhorn's theory before. This type generally used MBOs intuitively even at slow tempi. The "theorist" and the "intuitionist" types were identical in appearance. They differed only in the degree of awareness of their behaviour and in the underlying learning biographies. When a stronger increase in muscular tenseness led to a stronger decrease in the percentage of MBOs/HT under 60%, approaching the 'threshold' of 50% MBOs/HT associated with an increasing risk of acquiring severe PRMDs, the movement type could be characterized as *"stiffened."*

In contrast, the *"unleashed"* type showed extreme muscular control in a parallel leading technique at slower and medium tempi and gave up the muscular leading in favour of mass balancing oscillations only at tempi that did not allow for strong muscular control any longer with impunity. The "unleashed" type was able to switch between the movement qualities, whereas the *"inhibited"* type remained within the concept of "muscular leading" even at highest tempi and consistently played with MBOs below 50%.

The "soloist" type[6] started with quite high muscular control as well, but mass balancing oscillations began much earlier at a medium tempo and became very intense with the highest perfection at the freely chosen tempo. The MBOs declined at even faster tempi just as they did for the "theorist" and the "intuitionist" types. Only the attempt to play at an almost impossible tempo of 210 bpm finally led to stiffened muscles also in this profile.

## DISCUSSION

Previous studies suggested that the "desirable area" of MBOs/HT, associated with high expertise, ranges from 60% to 80% MBOs. The mean of 61.3% MBOs/HT seems comparatively low and close to the lower border of the assumed segment of high expertise. The "actual mean area" of orchestra violinists in this study formed, in fact, a wider segment from 43% to 78% MBOs/HT. Violinists playing with percentages of MBOs/HT lower than 50% were those who employed a "parallel leading technique" even at high tempi.

Looking more closely at the data behind the mean value of 61.3% MBOs/HT (see Figure 3) reveals a high variation (SD 11.4%). There is a pronounced scattering both in the minimum and the maximum directions in the main field of the subjects with a length of service between 18 and 32 years. There seems to be a division of the participants into pronounced users of MBOs and quasi non-users. The only recognizable association to this scattering in this study is the relation between the percentages of MBOs/HT and the suffering from a PRMD as well as the reported subjective impairments due to a PRMD. As PRMDs were reported more frequently among violinists playing with lower percentages of MBOs/HT, it might be that the area below the 50% threshold can be regarded as a "critical area" possibly associated with a higher risk of acquiring PRMDs. Correlation analysis and one-way ANOVA confirmed the importance of this relation.

The lower the percentages of MBOs/HT in violin bowing, the more frequently higher grades of subjective impairments were also reported due to a PRMD (see Figure 1). Thus, despite the fact that all participants served in their orchestras without noticeable absences,

---

[6] Measured MBO values in playing the musical task by a renowned German soloist. The data of the "soloist" type are real results of an additional subject for comparison purposes within this study. The "soloist" profile was found repeatedly among the orchestra violinists.

they could in fact be divided into healthier and less healthy orchestra musicians. This does not take into account the "healthy worker selection" bias of the test subjects, which is necessarily present in studies with participants who are active performers and thus exclude violinists who might be prematurely retired due to an occupational PRMD. On the other hand, participants in pain may have been more motivated to participate in the study, which could have had the opposite effect on the results.

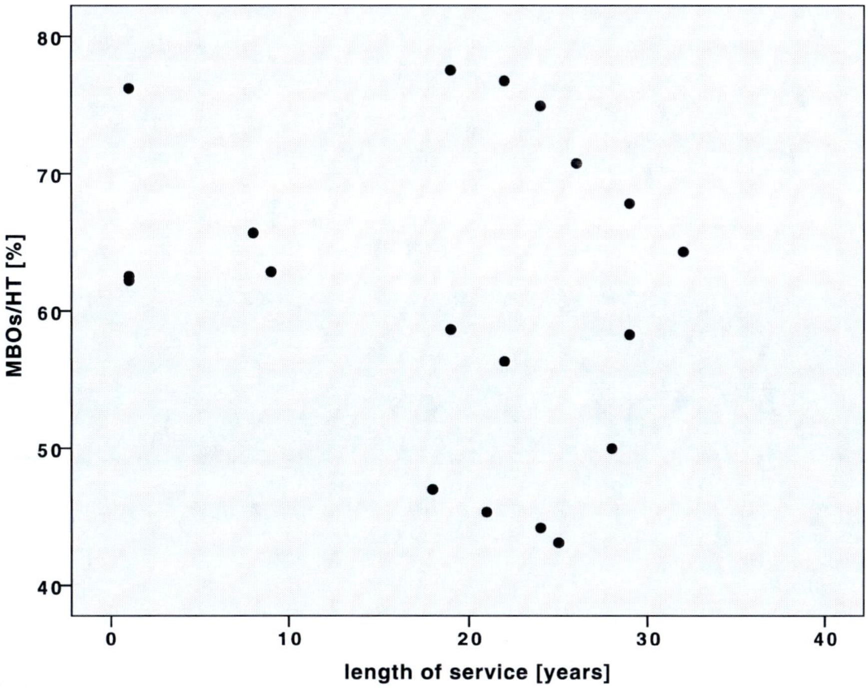


Figure 3. Scatterplot of the percentages of MBOs/HT of 19 orchestra violinists in the study compared to their "length of service" in the orchestra. The mean value of 61.3% MBOs/HT captured only slightly the actual reality of the values, which was rather characterized by strong deviations visible equally in both the minimum and the maximum directions (min. 43.1%, max. 77.5%, SD 11.4%). The ability to perform with high percentages of MBOs/HT was obviously neither a matter of age nor the length of service in orchestras among the participants in this study.

It may be questioned whether the simple task of playing a two-octaves scale with fast repetitive movements adequately depicts the reality of artistic violin performance or not. Looking at the following example of two participants in the study who spontaneously played J. S. Bach's prelude from Partita III in E major may be convincing. The artistic outcome was impeccable in both performances, but it was achieved with different movement qualities, just resulting in slightly different timbres and loudness. Which one may be preferred by an audience is probably a matter of taste and perception:

The percentage of MBOs just indicates the degree of approach to a maximized muscular economy. The fewer MBOs that are detected, the less free swinging of the levers happens and the bigger the influence of postural muscular activity is. If the percentage of MBOs increases above 50%, the bowing shows visible mass balancing oscillations, a "counter-swing-technique" (see Figure 5, where the upper line of the elbow oscillates in the opposite phase to the two other lines of the markers on the hand). If the percentage of MBOs falls below 50%, the tendency goes to a "parallel-leading-technique" (see Figure 4, where all lines are parallel). Thus, the mean value of 61.3% MBOs/HT for orchestra violinists represents a good tendency, but a mean value of 70% MBOs/HT with less pronounced deviations going below the 50% threshold seems much more desirable in the future, not only regarding the muscular economy but also the possible occupational risk of acquiring a PRMD.

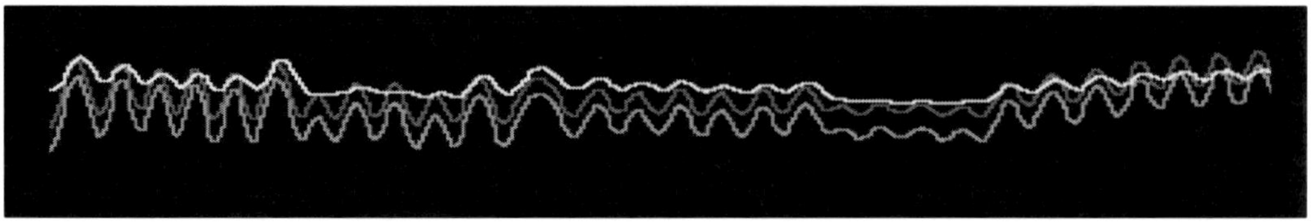

Figure 4. Measuring lines of a participant showing 58.3% MBOs/HT at high tempi (the "unleashed" type) but playing the prelude with approximately 20% MBOs at a medium tempo around 120 bpm.

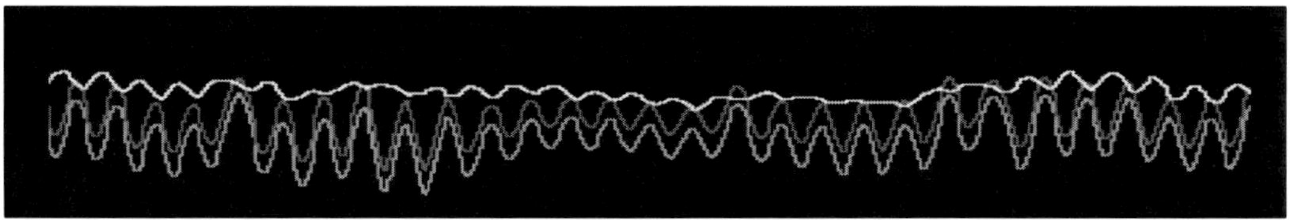

Figure 5. Measuring lines of a participant showing 58.7% MBOs/HT at high tempi (the "intuitionist" type, with tendency to the "stiffened" type) but playing the prelude with approximately 70% MBOs at a medium tempo around 120 bpm.

The negative correlation between the percentages of MBOs/HT and the suffering from a PRMD signifies only that there was a remarkable increase in the occurrences of PRMDs among violinists with low percentages of MBOs/HT, and those violinists without reported PRMDs had strikingly high percentages of MBOs/HT. However, this correlation says nothing about the possible causal connections. Perhaps low percentages of MBOs give rise to occupational PRMDs or they appear as a consequence of a PRMD. It seems likely that relationships exist in both possibilities. The shapes of the measuring lines give more concrete information about the possible character of a connection. If the line is very flat, strong muscle tensions inhibit a swinging of masses. But if the line shows a double size swing compared to "normal," this could indicate a compensatory movement pattern due to the absence of a functioning antagonist muscle. This was found in the case of focal dystonia in the bowing arm of one participant. The participant managed to play with the help of a movement comparable to an "easy trot" when riding a horse. This cut the percentage of MBOs in half, which might have been double in size before the onset of the disorder. Nevertheless, a decline in the percentage of MBOs may have been the trigger for the onset of the PRMD. Slight inhibitions of MBOs at high tempi can lead to stiffening muscles and blockages in the leverage system, accompanied by increasing psychological performance stress. This may end in a vicious circle and further declining percentages of MBOs/HT.

Even if there was no significant correlation detected in this study between the freely chosen tempo in the measured task and the percentage of MBOs, the percentage of MBOs is supposed to have influences on the feeling at ease at higher tempi. Violinists with completely different percentages of MBOs are frequently stand partners and have to cope with the same artistic tasks. For example, a participant in this study with 45% MBOs/HT often plays alongside a violinist with 77% MBOs/HT. The interview with the violinist, who showed only 45% MBOs/HT in the measurements, revealed the psychological performance stress connected with such differences: "My stand partner is always faster than me – a torturous role model!"

However, in this study the "freely chosen tempo" correlated only with "age" and "handedness." There was a marginally significant correlation between "age" and the reported "handedness," which may be explicable by the tendency to not force right-handedness today compared to earlier generations.

One can certainly criticize that the participants in this study were not very different in age. It would be interesting and enlightening to investigate the percentage of MBOs and the suffering from PRMDs of more violinists at different ages and to check if occupational PRMDs are really supposed to appear more frequently at a mid age and not at the beginning of a young career.

In this study, only five out of nineteen orchestra musicians could be considered "healthy musicians." This means that 73.7% of the investigated violinists were affected by considerable PRMDs. This corresponds to the results of Berque (2016), who documented a lifetime prevalence of 77.2% for PRMDs. In the study of Kok et al. (2015), the playing-related lifetime prevalence fell comparably between 62% and 93%. The results of this study matched the result of Aki & Yakut (2003) who found no significant correlation between "instrument playing time" and the extent of the complaints, which underlines the more decisive role of movement quality in comparison to movement quantity. Moreover, there were no significant correlations found with age, gender and the more general occupational stress while serving in the orchestra, a finding which contradicts some other studies (cf., e.g., Steinmetz et al. 2015) but goes partly along with the results of Kaufman-Cohen & Ratzon 2011, who concluded, "Gender turned out to be the only individual variable that predicted PRMD; a finding that contrasts with the published literature. Contrary to expectations, the psychosocial burden… was not found to predict musculoskeletal pain." (Kaufman-Cohen & Ratzon 2011, p. 90) They also found biomechanical risk factors to be one of the strongest statistical predictors for PRMDs among orchestral classical musicians.

Apart from investigating mass balancing oscillations as a central element of Eichhorn's leverage system, a suggestion for a future study would be to investigate possible relations between physical activities that prevent ailments and the percentage of MBOs. The highest probability for a significant result may offer the comparison between violinists practicing Dispokinesis[7] and those practicing Feldenkrais[8]. The hypothesis that the percentages of MBOs of Dispokinesis and Feldenkrais practitioners differ systematically is encouraged by clearly recognizable differences in the aims of these two methods of physical exercise. On the one hand, in the Feldenkrais method, muscles are exerted in "gentle, slow, repeated movements." The "attentive movement" as a "slow repetition" which "is believed to be necessary to impart a new habit and allow it to begin to feel normal"[9] is the opposite of swinging movements that are more or less as fast as gravity allows. On the other hand, in the Dispokinesis method, the focus is on enabling freely floating movements out of mostly

[7] cf. http://www.dispokinesis.de/en_gb/.

[8] cf. https://feldenkrais.com/.

[9] https://en.wikipedia.org/wiki/Feldenkrais_Method, 2017-09-24.

relaxed joint muscles thanks to tension in the lower abdomen muscles. This muscular disposition may facilitate the swinging in a leverage system.

Similarly, a comparison of jogging and high intensity training (HIT) could be interesting in the realms of sports. While in jogging the arms usually swing freely in the shoulder joints, the movements during HIT are explicitly not to be executed with swinging limbs but rather slowly under steady muscular control. A confounding reality is that musicians often exercise several sports and bodywork techniques in the same period of time. The most amusing but plausible statement of a violinist in the study was, "Playing the violin is 'good sports'!", which was underscored by his/her very impressive 76.7% MBOs/HT, being one of the healthy musicians.

It would have been interesting to evaluate an additional question in the interview concerning the subjective sensations of having "solo potential." Perhaps there would have been a significant correlation with the cluster highly above the average. It would also be desirable if more soloists appreciated the value of research in the future, despite their many concert engagements, and contributed to the empirical database. Their percentages of MBOs/HT and tempo-related profiles are necessary to consolidate the shape of the "soloist" type and the expertise paradigm connected to high percentages of MBOs. Unfortunately, only one renowned soloist could be persuaded to participate in this study. But interestingly, the two performance students from the pilot study 2008 (cf. Hasselbach 2010) have become highly admirable performers ten years later. One of these former students has expanded a solo career in the meanwhile, performing about thirty solo concerts per year. The second student has passed the entrance examinations for "Konzertexamen," the solo performance studies program at a renowned university of music, and was awarded the first prize in a German violin competition, indicating a promising and beautiful solo career in the future. This very small group of three soloists forms an informative database with a mean of 80% MBOs/HT, thus upholding the presumption of an expertise paradigm with a high level of expertise in the segment of 60% to 80% MBOs/HT, which should be proved further.

In the future, empirical longitudinal studies that take measurements at the beginning and at the end of studies at a music university and maybe also throughout the learning processes, or at the beginning of professional life and later on, may give better insights into the life-span development of individual tempo-related profiles of MBOs. The profile development may elucidate also the question whether a low percentage of MBOs gives rise to occupational PRMDs or is a consequence of a PRMD. Even if the chronological order as a criterion of causality still has to be investigated, the dose-response relationship as a criterion for causality is evidenced by the significant negative rank correlation between MBOs/HT and the severity of "subjective impairments due to a PRMD" (NRS-6). The economy of muscular effort in violin playing with a high use of MBOs may help to sustainably reduce severe PRMDs of violinists and likewise of violists, cellists and double bass players.

## Conclusion

Every violinist has a very personal learning biography and shows a resulting individual tempo-related profile of mass balancing oscillations. Biomechanically, a "perfectly economic" string instrumental bowing can be considered as a theoretical case. Nevertheless, a good "formula" in biomechanics can create a directive as in mathematics. Individual profiles can be visualized in a "field" of possibilities, compared to each other and localised in relation to a directive.

It seems plausible to assume that 100% use of MBOs is practically unachievable because human performances are expected to have limitations concerning perfection. An approach to the use of 90% MBOs may be the maximum of achievable muscular economy. On the other hand, an approach to 10% MBOs may represent the opposite end of the greatest possible muscular control of movements. Whether a violinist chooses more control or more economy is often – beyond the influence of predispositions, personalities and learning biographies – finally in most cases a question of tempo. If the requested tempo increases and approaches very fast repetitive movements, the effort of muscular control increases exponentially. Every violinist is presumed to have an individual critical point of tempo beyond which the body is likely to switch to "automatic economy operation," possibly to prevent the body's destruction or at least to minimize discomfort. But it is obviously possible to choose a much earlier point, even at the slowest tempi. This can be done without existential need to mitigate stress; it can be a conscious strategy employed just because these economic movements generally feel more comfortable and lead to artistically valuable sounds (cf. Hasselbach 2012, 2018). The ability to switch the movement quality and the ability to switch it early could be of great importance to avoid PRMDs.

On the one hand, low percentages of MBOs in repetitive movements at high tempi may contribute considerably to occupational PRMDs, to performance stress and to the possibility of premature termination of a musician's career, and, on the other hand, extremely high percentages of MBOs already at a medium fast tempo seem to be an outstanding feature of soloists. Therefore, this study recommends the training of mass balancing oscillations in the sense of Eichhorns's leverage system in string instrumental bowing both as a risk-reducing factor and as an artistically valuable factor, be it in the string instrumental education of children and students or in the preventive further education of professional orchestra violinists.

## REFERENCES

Aki, E. & Yakut, Y. (2003). Overuse syndrome and related problems in professional and student string players. *The Pain Clinic, volume 15* (number 3), 327-331.

Becker, H. & Rynar, D. (1929/1971). *Mechanik und Ästhetik des Violoncellspiels* [*Mechanics and aesthetics of violoncello playing*]. Second edition and photomechanical reprint of the edition 1929. Wien, Austria: Universal Edition.

Berque, P., Gray, H., McFadyen, A. (2016). Playing-related musculoskeletal problems among professional orchestra musicians in Scotland: A prevalence study using a validated instrument, the musculoskeletal pain intensity and interference questionnaire for musicians (MPIIQM). *Medical Problems of Performing Artists, volume 31* (number 2), 78-86.

Caldron, P. H., Calabrese, L. H., Clough, J. D., Lederman, R. J., Williams, G., Leatherman, J. (1986). A survey of musculoskeletal problems encountered in high-level musicians. *Medical Problems of Performing Artists, volume 1* (number 4), 136–139.

de Souza Moraes, G. F. & Antunes, A. P. (2012). Musculoskeletal disorders in professional violinists and violists: Systematic review, *Acta Ortopédica Brasileira, volume 20* (number 1), 43-47.

Fishbein, M., Middlestadt, S. E., Ottati, V., Straus, S. & Ellis, A. (1988). Medical problems among ICSOM musicians: Overview of a National Survey, *Medical Problems of Performing Artists, volume 3* (number 1), 1-8.

Fry, H. J. (1986). Incidence of overuse syndrome in the symphony orchestra, *Medical Problems of Performing Artists, volume* 1 (number 2), 51-55.

Hasselbach, J. v. (2018). Stressing "Unstressed:" Relaxed emphasis and lightness in the flow of movement: Understanding meter via physical assistance and supported metrical sound experience in bowed string instrumental performance, *Arts BioMechanics, Volume 2* (number 2), the following article in this issue.

Hasselbach, J. v. (2012). For 'lively' and 'intensely related' tones in bowed string instrumental performance: A response to Peter Röbke's introduction to "Das Musizieren und die Gefühle," *Arts BioMechanics, volume 1* (number 2), 115-130.

Hasselbach, J. v., Gruhn, W. & Gollhofer, A. (2011). Effects of training on mass balancing oscillations in the bowing of (pre)teen violin students: A quantitative micromotion study, *Arts BioMechanics, volume 1* (number 1), 1-14.

Hasselbach, J. v., Gruhn, W. & Gollhofer, A. (2010). Mass balancing oscillations: An indication of expertise in the bowing of violinists: A quantitative micromotion study, In S. M. Demorest, S. J. Morrison, P. S. Campbell (eds.), *Proceedings of the 11th International Conference on Music Perception and Cognition (ICMPC 11,* pp. 388-393). Seattle, USA: University of Washington.

Hasselbach, J. v. (2009). 100 Jahre Physiologic Turn in der Streichinstrumentalpädagogik: Eine Bestandsaufnahme. In N. Schläbitz (Ed.), *Interdisziplinarität als Herausforderung musikpädagogischer Forschung* [Years Physiologic Turn in String Instrument Education: An Inventory. In N. Schläbitz (Ed.), *Interdisciplinarity as a challenge of music education research*] (pp. 239-262). Essen, Germany: Die Blaue Eule.

Hopfer, M. (1941). Die Klanggestaltung auf Streichinstrumenten. *Das Naturgesetz der Tonansprache: kurze Einführung in "Die gestaltende Dynamik der Bogenmechanik" von August Eichhorn* [The sound design on stringed instruments. *The Natural Law of Tonal Speech: Brief Introduction to "The Forming Dynamics of Bow Mechanics" by August Eichhorn*]. Leipzig, Germany: Kistner and Siegel.

Kaufman-Cohen, Y. & Ratzon, N. Z. (2011). Correlation between risk factors and musculoskeletal disorders among classical musicians. *Occupational Medicine, Volume 61* (number 2), 90-95.

Kenny, D. T., Driscoll, T., Ackermann, B. J. (2016). Is playing in the pit really the pits?: Pain, strength, music performance anxiety, and workplace satisfaction in professional musicians in stage, pit, and combined stage/pit orchestras. *Medical Problems of Performing Artists, volume 31* (number 1), 1-7.

Kok, L. M., Huisstede, B. M. A., Voorn, V. M. A., Schoones, J. W., Nelissen, R. G. H. H. (2016). The occurrence of musculoskeletal complaints among professional musicians: A systematic review. *International Archives of Occupational and Environmental Health, volume 89*, 373-396.

Kopiez, R., Galley, N., & Lehmann, A. C. (2010). The relation between lateralisation, early start of training, and amount of practice in musicians: A contribution to the problem of handedness classification. *Laterality, volume 15* (number 4), 385-414.

Kopiez, R., Jabusch, H. C., Galley, N., Homann, H. C., Lehmann, A. C. & Altenmüller, E. (2012). No disadvantage for left-handed musicians: The relationship between handedness, felt constraints and performance-related skills in pianists and string players. *Psychology of Music, volume 40* (number 3), 357-384.

Leaver, R., Harris, E. C., Palmer, K. T. (2011). Musculoskeletal pain in elite professional musicians from British symphony orchestras. *Occupational Medicine, volume 61* (number 8), 549-55.

Morreau, A. (2002). *Emanuel Feuermann*. New Haven, USA: Yale University Press.

Oldfield, R. C. (1971). The assessment and analysis of handedness: The Edinburgh inventory. *Neuropsychologia, volume 9* (number 1), 97 113.

Sass, A. L. & Stoeving, P. (2017/1907). *Das Geheimnis auf der Violine und dem Cello in kurzer Zeit einen schönen, blühenden, das Orchester beherrschenden Ton und ein unfehlbares, rhythmisches Stakkato, durch direkte Ausbildung der betreffenden Muskelpartien zu bekommen* [*To get the secret on the violin and the cello in a short time a beautiful, blooming, orchestral sound and an infallible, rhythmic staccato, through direct training of the respective muscle parts*]. Reprint of the original edition 1907. Australia: Leopold Classic Library.

Shan, G. and Visentin, P. and Schultz, A. (2004). Multidimensional signal analysis as a means of better understanding factors associated with repetitive use in violin performance. *Medical Problems of Performing Artists, volume 19*, 129–139.

Sousa, C. M., Machado, J. P., Greten, H. J., Coimbra, D. (2016). Occupational diseases of professional orchestra musicians from Northern Portugal: A descriptive study. *Medical Problems of Performing Artists, volume 31* (number 1), 8-12.

Steinhausen, F. A. & Reuter, F. von (1903/1928). *Die Physiologie der Bogenführung auf den Streich-Instrumenten* [*The physiology of bowing on the string instruments*]. 5. Edition. Leipzig, Germany: Breitkopf and Härtel.

Steinmetz, A., Scheffer, I., Esmer, E., Delank, K. S., Peroz, I. (2015). Frequency, severity and predictors of playing-related musculoskeletal pain in professional orchestral musicians in Germany. *Clinical Rheumatology, volume 34* (number 5), 965-973.

Tapping. (2008). (Version 1.4) *Computer software for the measurement of handedness*. For description see http://shop.meditech.info/Software_8.html. Wedemark, Germany: Meditech.

Trendelenburg, W. (1925/1974). *Die natürlichen Grundlagen der Kunst des Streichinstrumentspiels* [*The natural foundations of the art of string instrument play*]. Unchanged photomechanical reprint of the edition 1925. Kassel, Germany: Hamecher.

Zaza, C. (1998). Playing-related musculoskeletal disorders in musicians: A systematic review of incidence and prevalence. *Canadian Medical Association Journal, volume 158* (number 8), 1019–1025.

# SECTION 2. DANCE

In: Advances in Arts Biomechanics
Editors: Gongbing Shan and Peter Visentin
ISBN: 978-1-53615-641-6


*Chapter 7*

# GENDER IN DANCE – AN INTERCULTURAL AND BIOMECHANICAL APPROACH

***Kulkānti Barboza***
Fachhochschule Münster, Germany

## INTRODUCTION

The following article focuses on a socio-cultural as well as biomechanical comparison of one of the many classical Indian dances called the "bharatanāṭyam" on the one hand and the classical French ballet on the other. My structural comparison of these two classical dance-systems is an attempt to make intercultural as well as intra-cultural understanding possible between the two. I will attempt this by taking into consideration the following:

- Two different societies (India and France)
- Two different historical eras, namely the 16th/17th and the 18th/19th century respectively

The research indicates that, to have a proper understanding of any dance system as well as the person of the dancer or indeed a particular dance-performance, a deeper knowledge of the socio-cultural background in which the dance is situated and from which it derives its inspiration, is necessary. Besides this, the analysis of Ballet will clearly show that cultural values are not static. Rather they are dynamic and change over the course of history.

Describing and interpreting cultural and epochal values of a dance-system should be the first step to approaching different body and movement cultures of our world and could be used as a beginning for understanding societies – both within and outside Western traditions.

The investigation undertaken in the current research focuses on one "universal" concept of movement, viz. the idea of "gender" which is found in all societies, but which can only be understand from the emic point of view, since each society has its own, specific cultural understanding of the performed gender roles – in social as well as artistic life.

## BIOMECHANICAL ANALYSIS

In addition to the ethnographic approach, a biomechanical analysis supports the investigation in order to see whether sociocultural, abstract ideas behind the art of dance are represented in body movement.

For this reason, a three-dimensional analysis of Indian as well as Ballet dance steps was undertaken, using a nine camera, VICON V8i system (VICON Motion Systems, Oxford Metrics Ltd., Oxford, England) at a recording speed of 120 Hz.

Each dance style was performed by both professional female and male dancers with ages ranging from 28 to 55.

42 spherical reflectors were placed on each dancer as follows:

a) Head
   - 4 reflectors

b) Torso
   - sternal end of the clavicle
   - xiphoid process of the sternum
   - C7 and T10 of the vertebral body (corpus vertebrae)
   - both scapulae
   - left and right of the iliac bone (anterior superior os ilium)
   - iliac bone (posterior superior ilium)

c) Upper extremities
   - right and left end of the spina scapulae/acromion (shoulder)
   - lateral side of the upper arm
   - lateral epicondyle
   - lateral side of lower arm
   - styloid process of ulna and radius
   - distal end of the third metacarpal bone

d) Lower extremities
   - left and right lateral side of the thighs
   - left and right lateral side of the shanks
   - joint head of the shinbone (lateral tibial condyles)
   - ankle (lateral malleolus)
   - distal end of the fifth metatarsal bone
   - heel bone (calcaneum)
   - upper side of the big toe (hallux)

## THE CONCEPT OF GENDER IN INDIAN DANCE (BHARATANĀṬYAM)

The Indian universe materializes itself on the basis of the two opposite but nevertheless together working forces, viz. the passive-male static energy (puruṣa) and the active-female dynamic energy (prakṛti). The eternal-female power, called śakti, embodies "the mystery of creation and the mystery of existence at the same time, the secret of all being and becoming, of the dying and in an incomprehensible way the re-born[1]" (Eliade 1985: 211, transl. K. B.). SHE it is, who creates and destroys, while the male (puruṣa) is motionless and an uncreative self-sufficient consciousness.

The polarizing pairs of Gods in the Indian Pantheon are arranged similarly. Each main deity (maheśvara) is only complete with its female counterpart, the goddess (devī) (cf. HKW 1992: Figure p. 160). Hence, an Indian deity is always a god-goddess. Together, as a united pair, they represent the primeval unfolding of the absolute Neutra (the puruṣa (n.)/the brahma (n.)). Thus, they are the first self-revealing incarnation of the Absolute, "imbued with the secret knowledge that, though seemingly two, they are fundamentally one" (Zimmer 1974: 137). The duality of existence and the duality of gender respectively, are understood as principles of consubstantiality and remain a cosmic play of the world of appearance (māyā). For Coomaraswamy (1991) this is a characteristic of Indian art per se, since it is always endeavoured to unite the male and female aspects in perfect form, viz. "... in all art there are ... masculine and feminine factors ... unified in perfect form" (ibid.: 112).

God brahmā appears with his consort sāvitrī, also called gāyatrī or sarasvatī, the mother of the holy knowledge of the Vedas (veda) and all art forms[2]. The goddess lakṣmī, the goddess of luck and wealth, completes with her female existence the deity viṣṇu. And, Lord śiva, without his female energy (śakti) - who appears like śiva himself in 108[3] or even 1008 forms - would not be even capable of bending a blade of grass: "Only when united with Shakti, has Shiva power to create, but without Her the Deva cannot even stir" (Sarabhai 1992: 24). Thus, śiva is sometimes depicted as an androgynous deity, as a half woman half man character (ardhanārīśvara) whose body has a right male and a left female side (cf. HKW 1992: Figure p. 140). The left represents the śakti, the dynamic cosmic energy and the right half śiva in his passive-static manifestation, i.e., śava. An alternative representation is that the front of the body is male while the back is female. It is only when śiva is united with his female counterpart that HE acquires HER energy to become the Lord of the universe. Without HER, HE is like a dead body. Thus, both become inseparable aspects of the same invisible principle, the ONE.

Naṭarāja[4], the king of dance, has a double sexual nature which is symbolised by his wearing of both male and female classified earrings as well as by the sun (sūrya, male) and the moon (candra, female) shown in his hair. "Here, the antagonistic principles unite to constitute a single organism, a paradox, representative of the intrinsically twofold nature of the onefold universe and its inhabitants, man" (Zimmer 1974: 216). Although completely different in action, both god and goddess radiate a silent oneness, he is and she "expresses his secret nature and unfolds his character" (ibid.: 199).

Hence, any erotic symbolism which we will encounter in classical Indian dance, must be understood in an allegoric way. Especially in the devotional[5] mode of expression, works with sensuous metaphors of a sexual relationship, represents the longing of humans for union with the divine, the whole.

---

[1] "...das Mysterium der Schöpfung und das Mysterium des Seins zugleich, das Geheimnis alles Seienden und Werdenden, Sterbenden und auf unfaßbare Weise Wiedergeborenen" (Eliade 1985: 211).

[2] In South-India sarasvatī is especially adored as the patron of all 64 arts (kalā).

[3] The consort of śiva appears in different forms, according to the specific aspect she represents, for example as kālī, durgā, gaurī, pārvatī, umā, satī, bhavānī, caṇḍi/caṇḍikā and cāmuṇḍā/cāmuṇḍī.

[4] In Barboza (in prep.) "Dancing body" the naṭarāja-Statue is depicted and interpreted in detail.

[5] See Barboza (in prep.) "Dancing realization".

The general goal is to realize and leave behind the dualism of the cosmic power as well as that of the sexes as appearances of matter, in order to immerse in the neutral realm of consciousness, which requires no more differentiation, because it is beyond all manifestation: "... and only when the distinction and limitation of male and female is overstepped and the impersonal brahman realized can liberation occur" (Stutley 1977: 18).

In analogy to the cosmos and the divine world, the south Indian dance style bharatanāṭyam has likewise androgynous characteristics. Apart from the classification system of dramaturgical (nāṭya), abstract (nṛtta) and concrete (nṛtya) representation technique, there are two categories concerning the way of dancing:

a) powerful-male[6], now and then even devastating (tāṇḍava)
b) tender-female (sukumāra prayoga/ sukumāra nṛtya/lāsya) (Vatsyayan 1977: 25 f.)

Although these classifications contradict the above mentioned right-left-opposition of the body and its male-female attribution (in which the female represents the dynamic and the male the static power), it shows a clear male-female-opposition. In addition to this, it is interesting to note that the term "tāṇḍava" is not solely male, it is at the same time a superordinated, gender-neutral category: "The Nāṭyaśāstra uses the word tāṇḍava as a generic term which cannot be interpreted as denoting vigorous dancing or as one performed by men alone" (ibid.: 25). As such, it combines "recaka" (circular movements of a limb), the aṅgahāra (sequence of movements) and the piṇḍī (grouping) with song and instrumental music" (ibid.: 25). This means that, besides the two polarizing genders, there is a "third gender," - the Neutra, - in which sexual dualism withdraws itself or is rather "neutralized."

According to its mythological origin, Lord śiva's dance was passed on from the deity to his devotee taṇḍu as was the female aspect of dance from the goddess pārvatī to ūṣā, the goddess of dawn, in order to become accessible to mankind (Bharata n. y.: IV, 19-264).

The two following examples illustrate how the androgynous idea of south Indian dance enables a dancer to assume (simultaneously or with some delay) both gender roles – independent of the biological sex of the dancer:

a) The right-left-opposition of body movement represents not only the completing forces of activity and passivity but represents analogous a female-male-opposition, since the active state is classified as female and the passive one as male. Likewise, the triangle constellation (trikona) of the basic dance position (ardhamaṇḍali) reflects the śiva-śakti, the cosmic principle of male passivity and female activity.[7]

Since both aspects are always depicted in basic dance positions simultaneously, one can reasonably hold the view that in Indian dance the concept of the person is of bisexual nature: the dancing person is, like the deity śiva, both man and woman at the same time.

b) Study of Indian dance shows that each student, regardless of his/her biological sex, learns the same basic steps and the same dance repertoire. As further evidence, the Tamil word for temple dancer, dēvaraḍiyal, covers both female and male servants of

[6] The fact that tāṇḍava is classified as active-male and thus contradicts the principle of the active female energy śakti, makes one to presume a syncretistic influence of Arian culture. In spite of its contradiction, both categories exist in Indian dance in harmonious coexistence.

[7] For further details see: Barboza (in prep.) "Dancing body".

God (Sadasivan 1993: 1; footnote 2). Unlike French ballet, there is no gender specific movement pattern for male and female dancers. Rather, special movements and gestures are classified as male or female and both must be mastered by each dancer.

Biomechanical data gathered during performances of Indian dance and Western classical ballet reflect very clearly differences between Indian and European dance. Both the movement patterns and the movement radia are almost alike in Indian dancers. Differences are rather more of quantity (like 90° flexion in one person and 95° in the other) than of structural nature (quality). Whereas in some cases – in particular the dance steps battement tendu, grand battement jeté and pas de bourrée – the movement pattern of the male ballet dancer differs from the sequences of the female counterpart in quality. Here, the attribution of gender does not merely concentrate on the use of strength (e.g., different athletic leaps). In parts, the entire structure of dance steps is differently oriented in order to separate both gender performances clearly. This distinction is evident as early as Feuillet's choreographic compositions of the early 18th century which identify several gender specific steps. For instance, the "folie d'Espagne pour femme" is designated specifically for a female dancer while the "gigue pour homme," "entrée pour homme" and "entrée grave pour homme" are identified for males (Feuillet 1700: 33-66).

As shown in the figures below, the gender specificity of ballet steps on the basis of different parameters of the step sequence called battement tendu is seen:

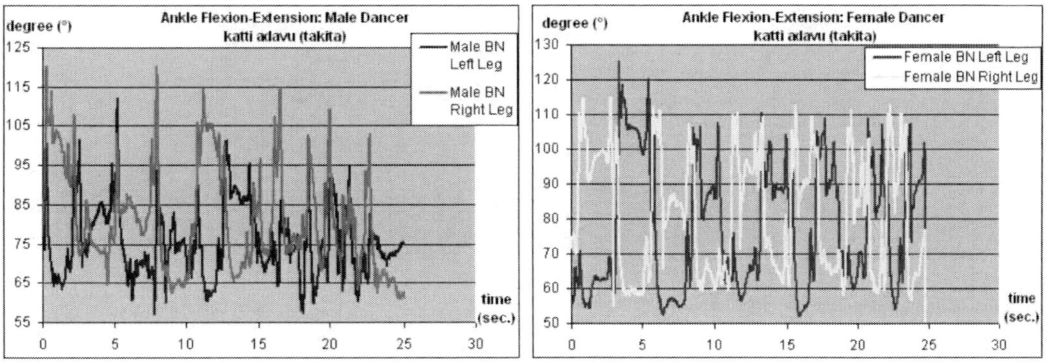


Figures 1 and 2. Ankle flexion/extension, katti aḍavu, M/F.

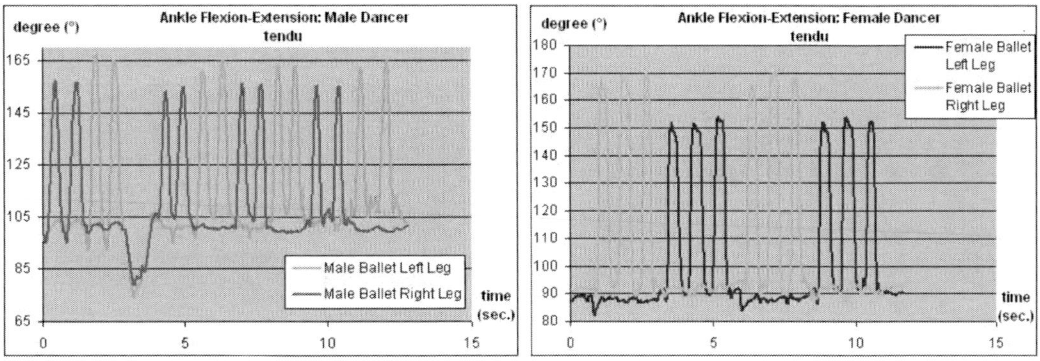


Figures 3 and 4. Ankle flexion/extension, tendu, M/F.

Except for the spatial range of the steps along the X-Y axis (Figure 17/18: heel movement (right)) and the flexion/extension of the shoulder, all Indian dance steps are similar, whereas there are remarkable differences between the genders in ballet in the execution of the battement tendu. Only in the arm movement of ballet (shoulder flexion, elbow flexion, wrist flex/extension, abd/adduction, rotation) there seems to be a convergence between both genders. The lesser variation between the genders is possibly due to the fact that there are fewer variations for the arm movement in general for ballet than for Indian dance. In addition to this, the emphasis in the battement tendu is on the leg work.

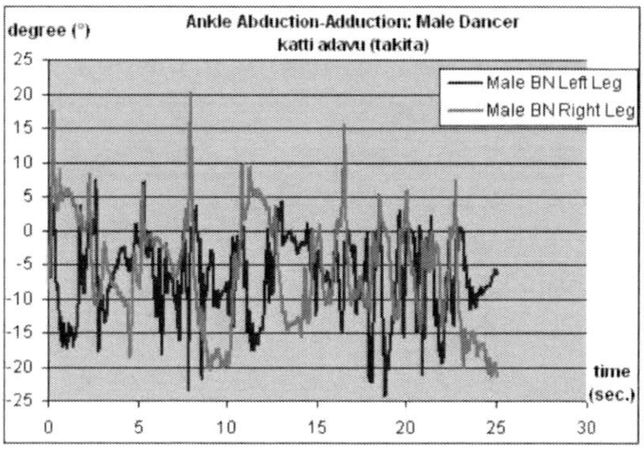


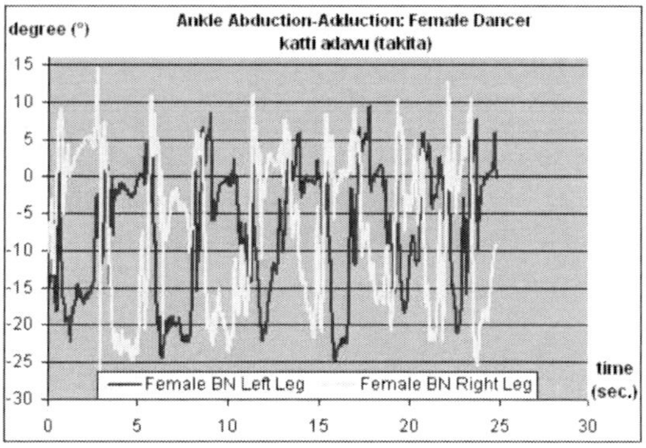


Figures 5 and 6. Ankle abduction/adduction, katti aḍavu, M/F.

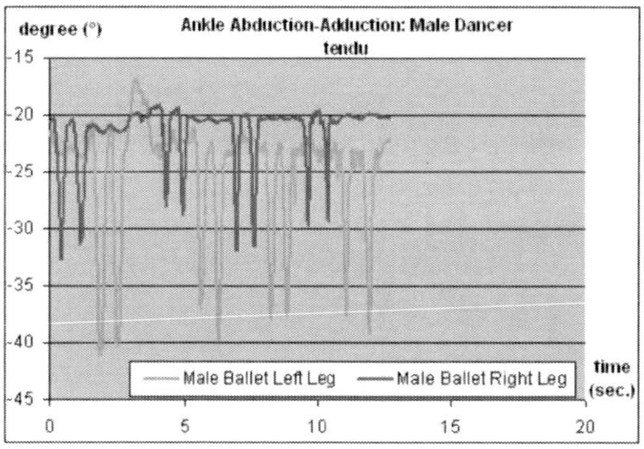


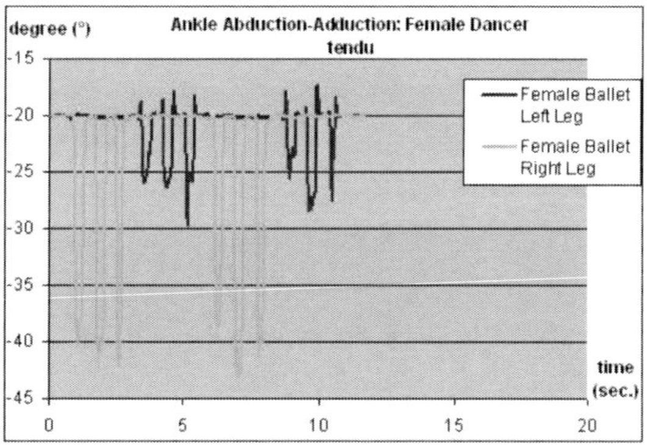


Figures 7 and 8. Ankle abduction/adduction, tendu, M/F.

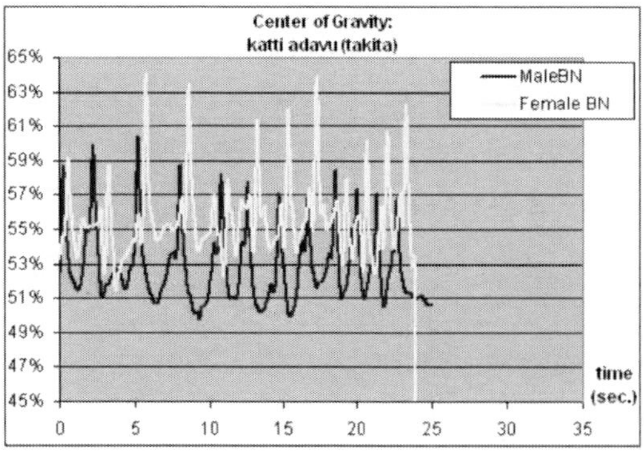


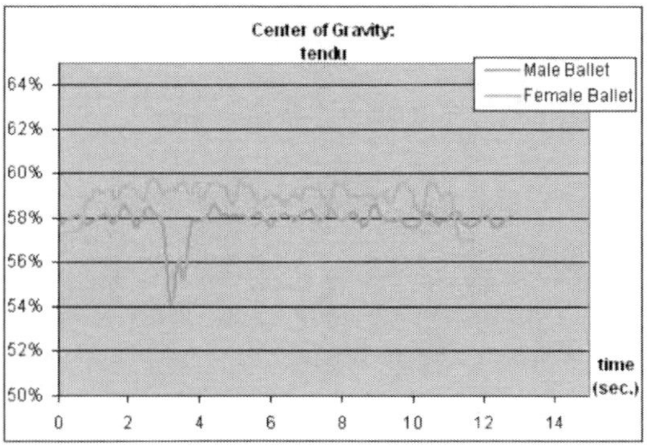


Figure 9 and 10. COG, katti aḍavu, M/F. COG, tendu, M/F.

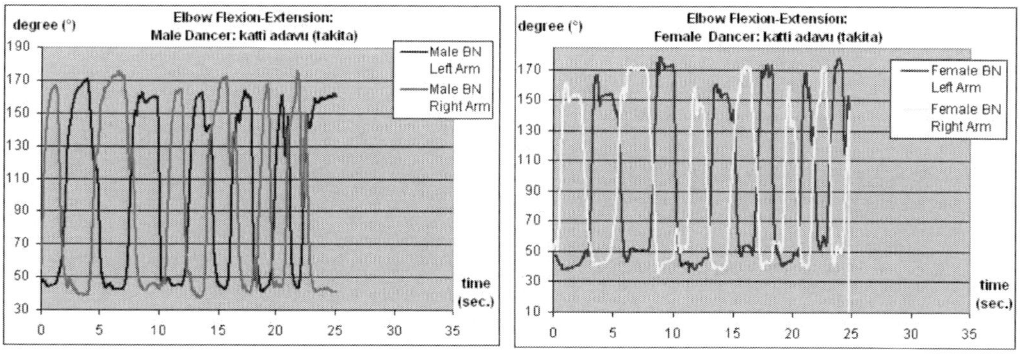


Figures 11 and 12. Elbow flexion/extension, katti aḍavu, M/F.

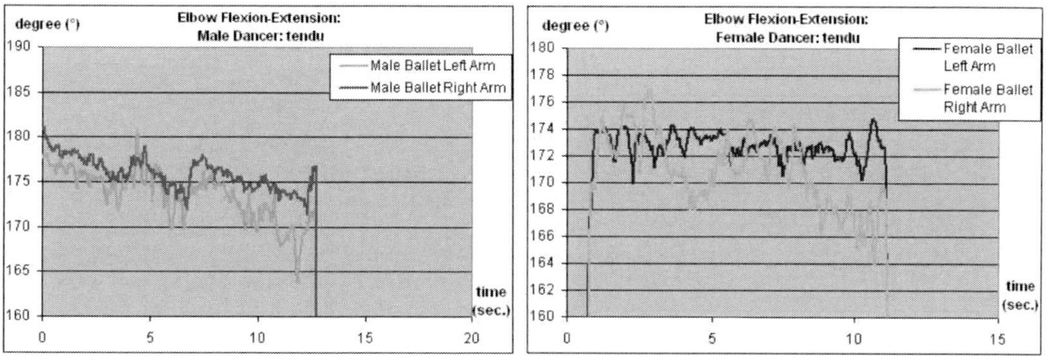


Figures 13 and 14. Elbow flexion/extension, tendu, M/F.

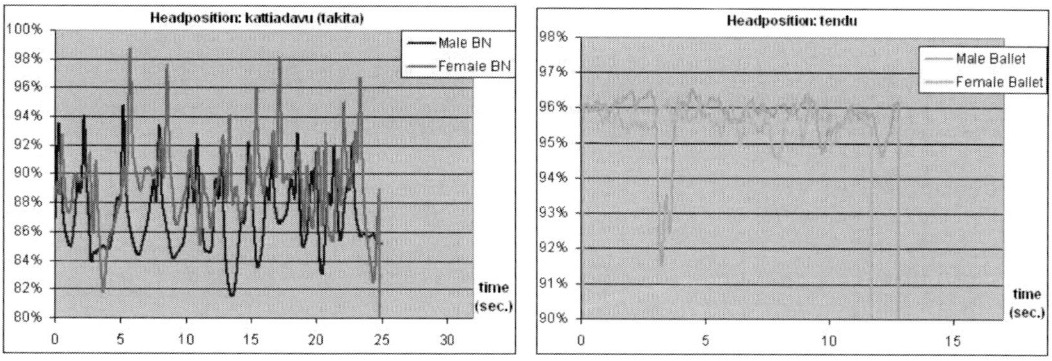


Figure 15 and 16. Percentage head height katti aḍavu, M/F. Percentage head height, tendu, M/F.

Typically, in ballet there is not only a clear separation of the sexes, but the distribution of the gender roles depends on the biological sex of the dancer. This is not the case with Indian dance. For typical Indian dance movements the idea of being a man or a woman unfolds during the act of dancing. The dancers are active participants in a role performance. They put into action simultaneously (e.g., as per the triangle constellation (trikona) of the basic dance position) as well as consecutively opposite aspects of gender performance. Thus, Indian dance becomes a prime example of constructing gender identities and hence of "doing gender" - completely independent of a biological conception of sex.

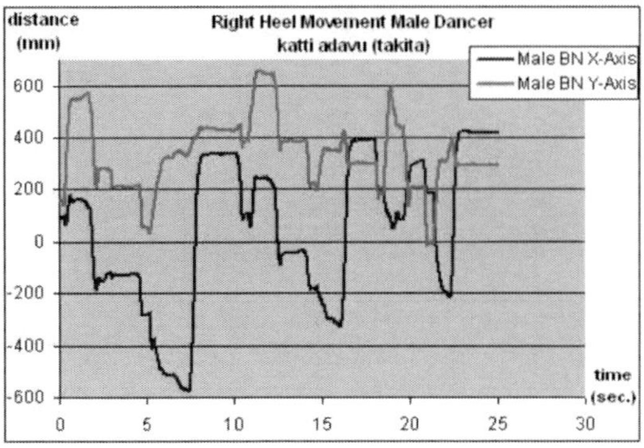


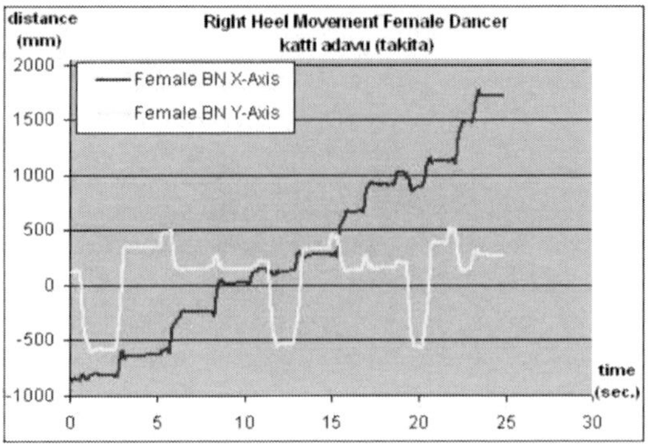


Figures 17 and 18. Heel movement (right), katti aḍavu, M/F.

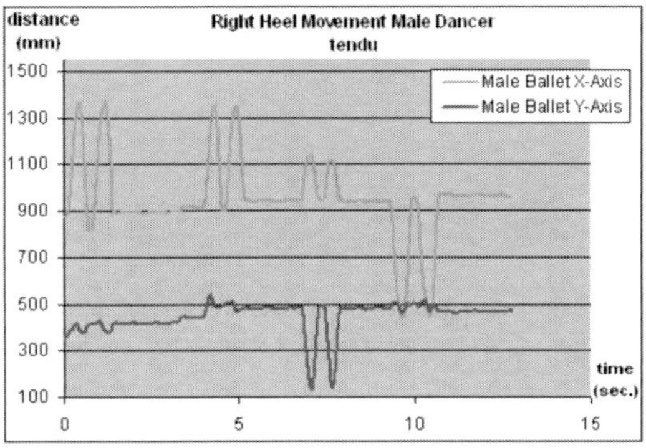


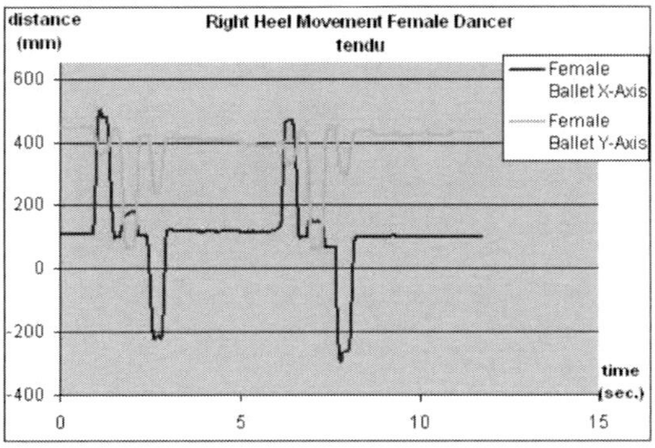


Figures 19 and 20. Heel movement (right), tendu, M/F.

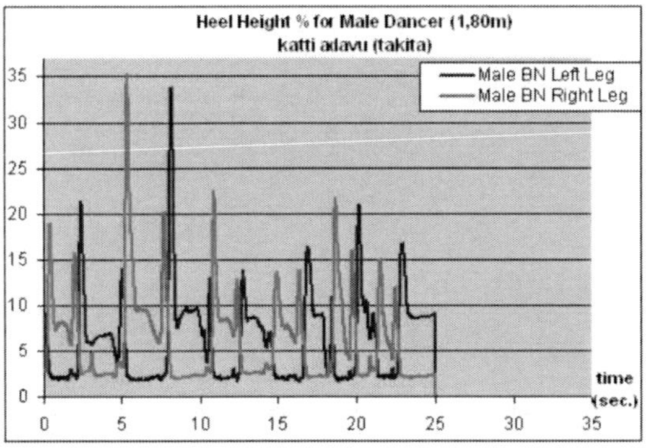


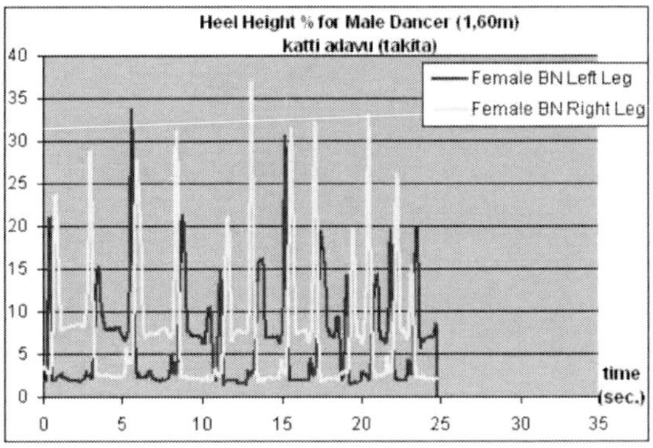


Figures 21 and 22. Percentage[8] heel height, katti aḍavu, M/F.

[8] The percentage of the heel height is related to the body height of each person (body height = 100%).

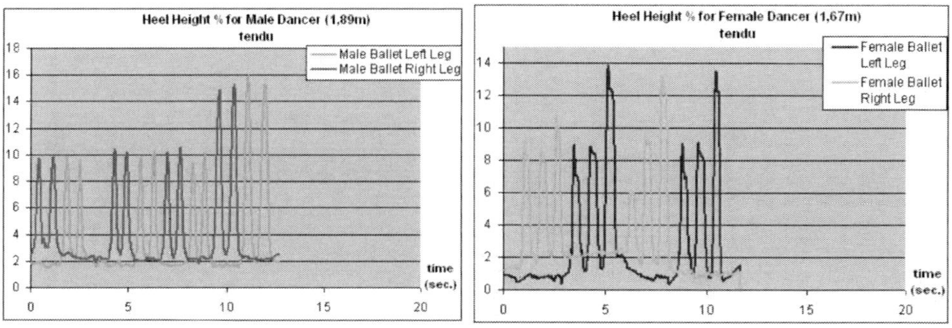


Figures 23 and 24. Percentage heel height, tendu, M/F.

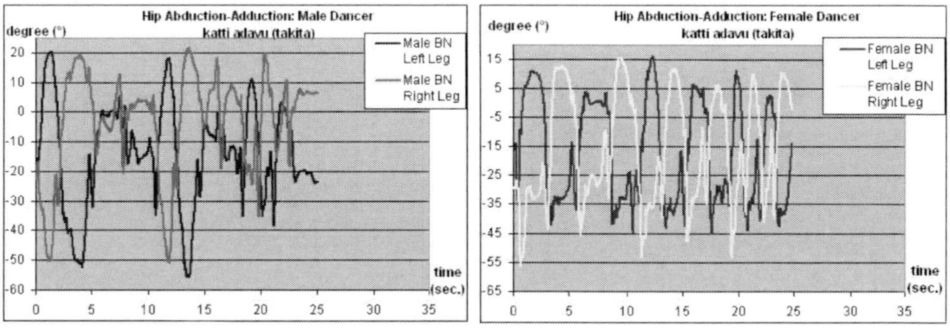


Figures 25 and 26. Hip abduction/adduction, katti aḍavu, M/F.

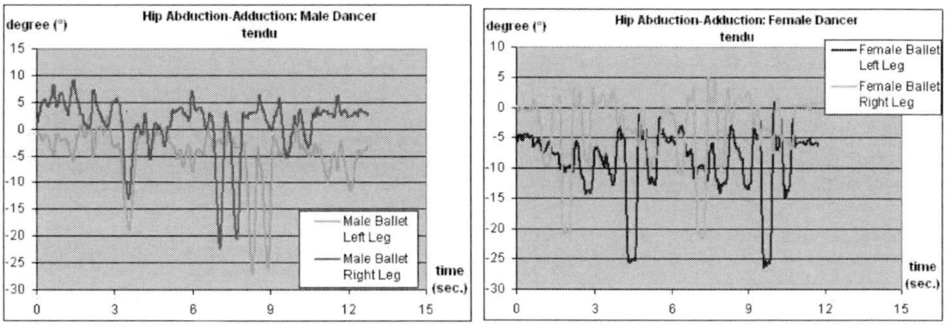


Figures 27 and 28. Hip abduction/adduction, tendu, M/F.

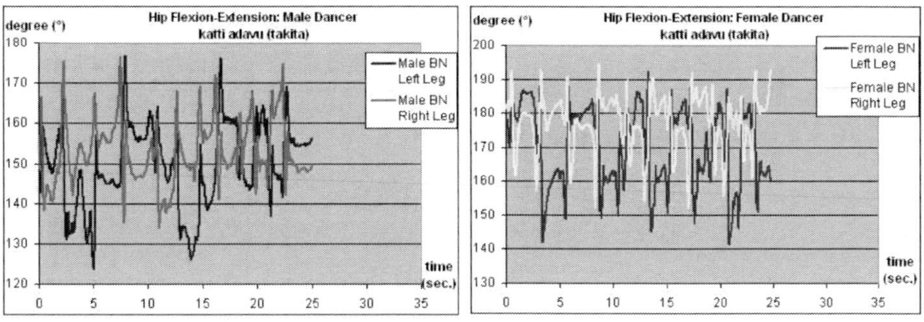


Figures 29 and 30. Hip flexion/extension, katti aḍavu, M/F.

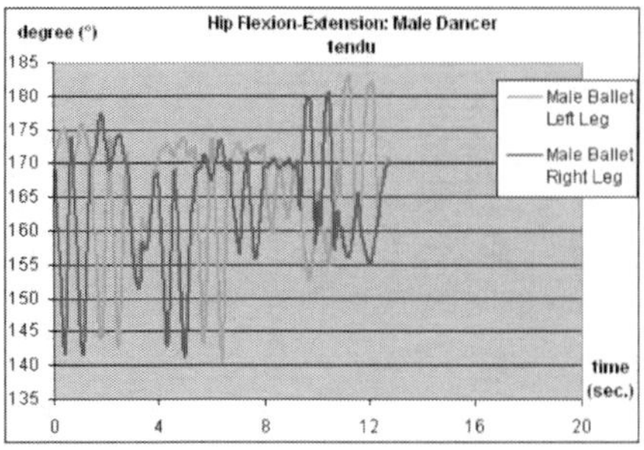


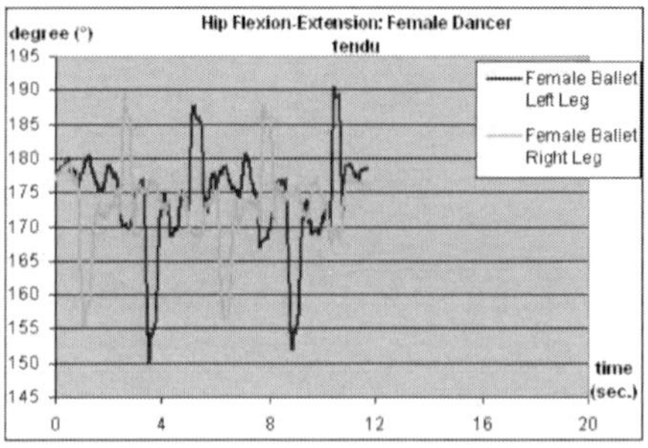


Figures 31 and 32. Hip flexion/extension, tendu, M/F.

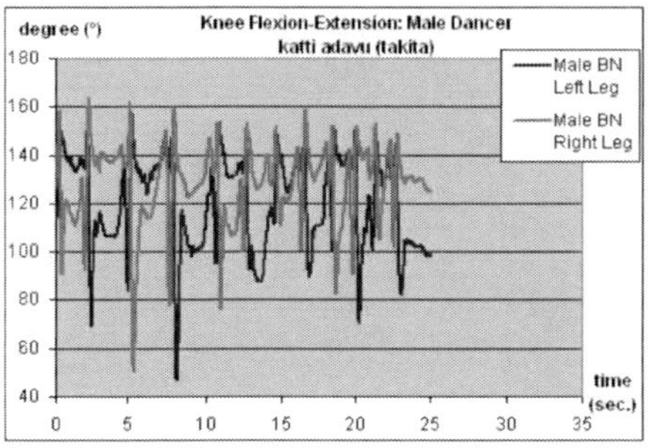


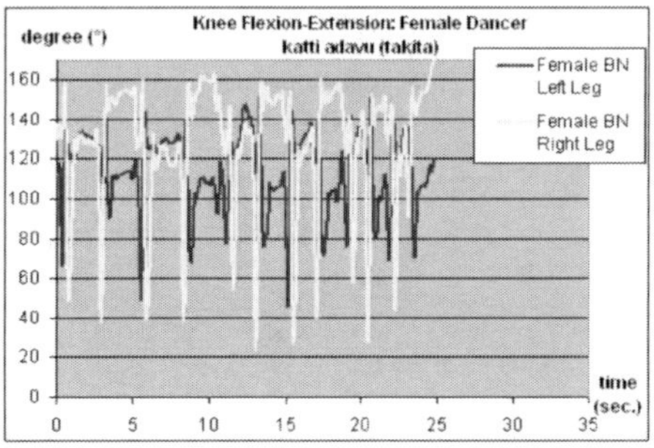


Figures 33 and 34. Knee flexion/extension, katti aḍavu, M/F.

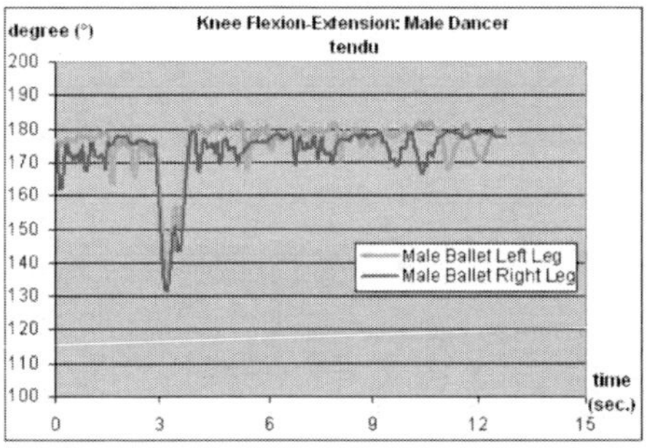


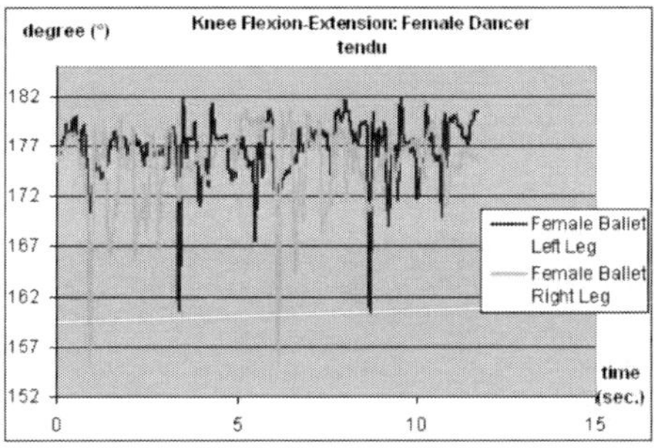


Figures 35 and 36. Knee flexion/extension, tendu, M/F.

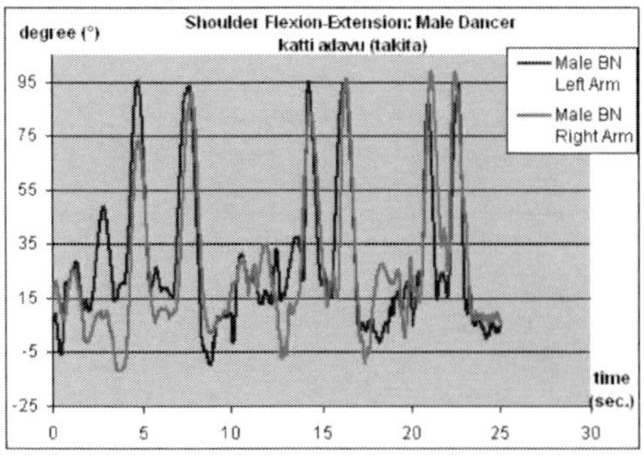


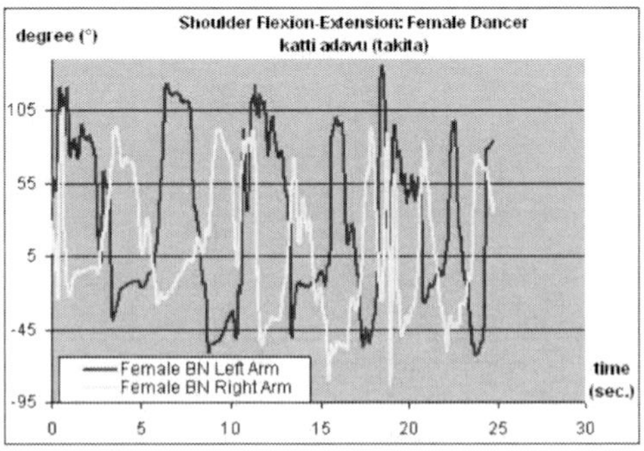


Figures 37 and 38. Shoulder flexion/extension, katti aḍavu, M/F.

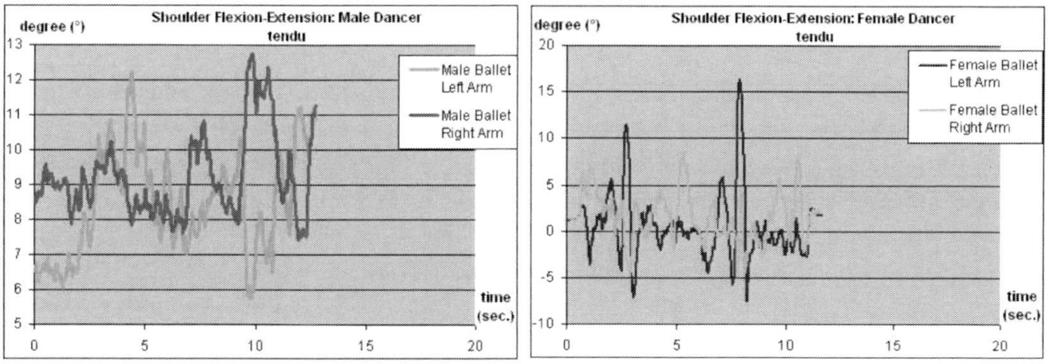


Figures 39 and 40. Shoulder flexion/extension, tendu, M/F.

In summary, the concept of gender in Indian dance may be understood as follows:

a) There are four categories of gender in Indian dance:
   - female
   - male
   - female-male (at the same time)
   - neuter (neither female nor male)
b) Differentiation is independent of the biological sex of the dancing person. Everyone must be able to represent all gender categories. The dance becomes thus a gender performance ("doing gender")
c) Right-left-opposition of the movement visualizes not only the cosmic pair of active and passive forces, but also that of the male and female one.
d) Female and male aspects do not necessarily exclude each other, but can be performed at the same time
e) Biomechanical data of the movement of the Indian dancers do not indicate a gender specificity, those of ballet do
f) The final goal of Indian dance is to overcome the female-male opposition which exists only in the cosmic world of māyā: pure consciousness is neutral (indicated in the Sanskrit-grammar as neuter: e.g., the puruṣa, the brahman or the tāṇḍava)

## Comparison of the Gender Concept in French Ballet and Indian Dance

As early as "Apologie de la Danse" by François de Lauze (1623) French court dance differentiates the behaviours of women and men (cf. Blomkvist 1996: 37 f.) Since then, it is possible to observe that the conception of gender and its role allocation has changed considerably over the course of time. For this reason, the following analysis examines gender categories one by one and, according to their specific epochal background, in order to uncover the socio-historical distinctions and draw some generally valid intercultural perspectives with Indian dance.

## Renaissance (14th -16th Century)

During the Renaissance, unmarried women were subject to social and behavioural restrictions (Jung 2001: 346). For instance, a woman might attend a dance event only during daylight. Certainly, she would not be allowed to go home at night in company of a man (ibid.: 344). Vosz (1869) mentions that a male might escort a single woman home, but the unmarried woman would have to walk in front of the man or 'pure and chaste' beside the gentleman. These conventions may not be intentionally directed toward the suppression of women by men, but could be explained by the fact that unmarried women were thought of as very valuable "gifts" of the society at that time and marriage was – like Jung (2001) documents – the most important transition ritual of the Early Modern Age (ibid.: 144-195).

Burckhardt (1976) points out the fact that women and men were of equal value during the Renaissance: "To understand the higher sociability of the Renaissance it is finally substantial to know that women and men were equally respected[9]" (ibid.: 368, transl. K. B.). The value of the woman, for example, shows up in social life that in the upper stratum of the society the education of the woman was generally the same as for the man (Burckhardt 1976: 369).

We find a similar accentuation of the female roles in ballet at that time, even if a shift is indicated in the Baroque conception of gender. The ballet drama[10] "Circé," premiered in 1581, deals with Circe, the Homeric sorceress and daughter of the sun God Helios and of Parse. According to Greek mythology strangers who, – like the companions of Ulysses, – enter Circe's island Aiaia are transformed into animals. Only Ulysses is able to protect himself against her magic power (Ranke-Graves 1989: 678 f.). In the ballet drama of the Renaissance it is Jupiter, the Roman father of Heaven, who defeats the enchantress and therefore releases[11] all those spellbound ones (Pasi 1980: 43 ff.).

The performed dance drama reflects:

a) the powerful, supernatural strength of the female one as well as
b) the subjugation of this female strength to the kingship or to a patriarchal social order

Regarding this ballet drama, the following facts show it to represent an "idealization of the feminine" (Klein 1992: 110) or at least a high value of the female potential:

- it is dedicated to the queen (la Reine)
- the main role[12] of Circe is danced by a woman
- the queen of France as well as the princess of Lorraine participated in the ballet as naiads[13]
- the ending of the grand ballet is initiated under the guidance of the queen
- during the different, geometrically arranged dance formations the queen always remains as the apex or the centre of the performing group (Pasi 1980: 43)

---

9 "Zum Verständnis der höheren Geselligkeit der Renaissance ist endlich wesentlich zu wissen, daß das Weib dem Manne gleich geachtet wurde" (Burckhardt 1976: 368).

10 "Ballet comique de la Reine".

11 See also Barboza (in prep.) "Dancing action".

12 Only 100 years later, with the approach of the prima ballerina Mlle. La Fontaine in the first ballet opera "Le triomphe de l'amour", the female dancers as main performers return to the domain of the classical dance (Klein 1992: 110-112).

13 Naiads are nymphs of the river.

With the appearance of Jupiter or rather the king, the creative aspect of the female gender is captured in order to be at man's disposal. Klein (1992) speaks of the "subjection of nature as well as of female sensuality[14]" (ibid.: 110, transl. K. B.). From now on the woman is subordinated not only to the man, but is also instrumentalized by him. The independent realm of the goddess is incorporated by the male deity. Since Circe is not killed, she and her productive as well as reproductive power continues to live, but not in an independent "female" existence, but for the purpose of the male ruling system. This modified Circe myth – which Cassirer (1990) would refer to the 'law of metamorphosis[15]' (ibid.: 130), and Blumenberg (1996) would discuss it in the context of "the hereditary persistence of outliving all passing historical changes[16]" (ibid.: 68, transl. K. B.) – stands for the transition from an amorphous woman power to an institutionalized male hegemony[17] and legitimizes or reinforces by use of ballet performances the royal rule structure of the Early Modern Age.

Considered together, both aspects reflect an ambivalent conceptualization of the female which we find again particularly in Romantic ballet and which stems from a Christian double conception of the woman. For example, in the Bible the woman personifies the source of all sin as well as of spotless innocence: Eve, the epitome of sexuality, seduction and sin opposes irreconcilably Virgin Mary in her asexuality and purity.

## Baroque (17th Century)

During the Baroque era from 1600 until 1680, the court ballet[18] is predominantly danced by noble men who also take over female roles which are called "rôle travesty" or "en travesty(e)", because they used to perform in ladies' dresses (Prest 2006: 79 ff.). Women – if at all – sporadically only are involved in a kind of finale (Klein 1992: 110; Liechtenhan 1983: Figure p. 57). The famous "Ballet Royal de la nuit" (1653) represents such a work par excellence which was performed by male dancers only, including the important female roles of "Venus" and "Agora" (Ballard 1653).

After the 1660's, as seen in the "Ballet d' Hercule amoureux", composed by Jean-Baptiste Lully in 1662, very few aristocratic ladies were allowed for major roles and later on professional women, too – like Mlle de Lafontaine's debut in 1681 in "Le Triomphe de l'Amour" (Sator 2003; Prest 2006: 103 ff.). The forthcoming of female dancers might have been influenced by the foundation of the "Académie Royale de danse" in 1661 which made ballet dance more professional and too athletic for most of the royal dancers. This was supported by Raoul-Auger Feuillet's and Pierre Beauchamp's publication "Chorégraphie" in 1700 which included the active participation of female dancers.

In contrast to the Renaissance ballet, a strict and conservative separation between the male and female sphere is found. Moreover, the idea of gender – similar to the Indian concept– becomes a "performative act" (Butler 1991: 198-208) in which the gender role can

[14] "Unterwerfung der Natur wie auch der weiblichen Sinnlichkeit" (Klein 1992: 110).
[15] "Gesetz der Metamorphose" (Cassirer 1990: 130).
[16] "...hereditären Hartnäckigkeit des Mitgehens durch die Geschichte" (Blumenberg 1996: 68).
[17] See for instance Lenz/Luig (1990), Lenz (1997), Meier-Seethaler (1993).
[18] "Ballet de cour".

be allocated independently of the biological sex. The 'biological[19]' man can dance either a male or a female role. Yet, there is neither the Indian constellation of the simultaneousness of gender (man-woman) nor the one of the Neutra. In addition to this, it is only the man who can take over both gender roles, whereas in Indian dance any dancer – irrespective of their biological sex – is enabled to participate in the play of double gender role.

Also the social background is different between these two cultural dance archetypes. The Baroque court dance occurs in concordance with a restricted code of conduct for women (subordination, restraint) and therefore does not allow them to take an active part in the ballet. A male dancer takes over the female role due to the social conventions, but not because it is a generally accepted socio-cultural concept. Instead, the male dancers jump over their boundaries in the true sense of the word: On the one hand they need to fulfil the political function of the dance, i.e., the demonstration and stabilisation of the hierarchical order at the royal court (which stands in place of the society) even while the Thirty Years' War raged over 'Europe'. On the other hand, a gender specificity arises which puts social focus on the dynamic, expanding and controlling man (Klein 1992: 110 ff.). It is no coincidence that dancing becomes a royal discipline at the royal dance academy, just as riding and fencing became social disciplines, in order to be of military use (Liechtenhan 1983: 38).

At the end of the Baroque era, gender relations changes, as may be seen in the ballet opera[20] "Le triomphe de l'amour[21]." This work was premiered at the Paris Opera in 1681. In it, the first professional prima ballerina, Mlle de Lafontaine (La Fontaine) dances a pas de deux[22]. The emancipation of the female sex is in the offing and it becomes evident in the epoch of Rococo – at least in the courtly world.

## Rococo (18th Century)

During the 18th century ballet undergoes fundamental changes especially due to the papers "Lettres sur la danse et sur les ballets[23]" by Jean Georges Noverre (1760) which demand a 'natural' way of expression and which enable particularly the female ballerinas extended movement patterns. Masks, wigs and movement restricting costumes are dispensed with. Great female dancers such as Marie Sallé[24] and Maria-Anna de Cupis de Camargo[25] enriched the royal theatre by their natural facial expression as well as by their dynamic leaps (Liechtenhan 1983: 51 ff.). Voltaire writes about the sultry dancer Camargo that she dances 'like a man' (ibid., in: Sorell 1985: 157). She and the many following prima ballerinas initiate not only the emancipation of the woman on the stage but snatch away from the male dancers their supremacy for the next two centuries.

This behaviour corresponds to the royal social life of the Rococo era which was considerably more progressive than the world of the middle-class. The ascending bourgeoisie

19 According to Butler (1991) even the so called "biological sex" has always been a culturally constructed gender identity. But I keep both the categories of "sex" and "gender" in order to differentiate between the sex of the "body" and that of the social "role".

20 "Opéra ballet".

21 'The triumph of love'.

22 "Step/dance for two", a duet in which a female and a male ballet dancer perform together.

23 'Letters on Dancing and Ballets'.

24 Marie Sallé lived from 1707–1756 (Günther 1959: 116).

25 Maria-Anna de Cupis de Camargo lived from 1710–1770 (Günther 1959: 116).

justifies the incapacitation of the woman due to her biological determination and they are – like everything ruled by "Nature" – socially excluded and suppressed. Even in the doctrine of ethics and rights by Johann Gottlieb Fichte[26] (1971) of the 18th century it is still written that the dignity of a married woman depends on the fact "that she is totally subjected to her husband[27]" (ibid.; transl. K. B.). This was justified not only in terms of natural law but also by the reasonable will of a woman, because "she is subjected by her own continuing, indispensable morality causing her desire to be subjected. She could take back her freedom, if she wanted to, but this is the exact point; to have a good sense she does not want it[28]" (ibid.: 345, transl. K. B.).

In contrast to this, a kind of gender equality appears in the royal world which partly opens up an hitherto unimagined world of erotic and mental realm of power: "The mental freedom of the 'République des Lettres' and the erotic scope for development which the 18th century enabled, contributed substantially to the liberation of women[29]" (Heyden-Rynsch 2004: 116, transl. K. B.). Even the cult of mistress is by far more than only an aristocratic form of sexual exploitation, since the mistress of the king is at the same time an influential woman at the court who influences important political decisions: "The coming to the fore of the mighty mistresses and the elegant courtesan as well as the etiquettes and the decisions of their gentlemen shaped the fancy and the self-confidence of the woman of high rank who had in former times disgustingly turned away from these notorious ladies[30]" (Heyden-Rynsch 2004: 116, transl. K. B.).

## Romantic (19th Century)

After the French Revolution and in concordance with the emergence of modern nation states, Romantic ballet develops which is tailor-made for the ballerina (Liechtenhan 1983: 74). The triumphal march of the once so weak and completely excluded sex is not to be stopped any more. The names of dancers like Fanny Elßler, Marie Taglioni, Fanny Cerrito and Carlotta Grisi became real cult figures. About Taglioni it is said that the U. S. Congress shifted a meeting in order not to miss her performance. They were worshipped by individuals from all parts of society, exquisite lady fashion developed and their personal items became collectables, e.g., Cerrito's bath water was sold at allegedly high prices (Zacharias 1962: 58-61).

From the dancing point of view as well, the prima ballerina dominates the man who is almost acting like a supernumerary, if he holds, carries and catches the dancer, in order to

---

[26] The German philosopher Johann Gottlieb Fichte lived from 1762-1814.

[27] "...dass sie ihrem Manne ganz unterworfen sey und scheine" (Fichte 1971: 345).

[28] "...sie ist unterworfen durch ihren eigenen fortdauernden nothwendigen und ihre Moralität bedingenden Wunsch, unterworfen zu seyn. Sie dürfte wohl ihre Freiheit zurücknehmen, wenn sie *wollte*; aber gerade hier liegt es; sie kann es vernünftigerweise nicht *wollen*" (Fichte 1971: 345).

[29] "Die geistige Freiheit der »République des Lettres« und die erotischen Freiräume, die das 18. Jahrhundert ermöglichte, trugen wesentlich zu[r] Emanzipation der Frau bei" (Heyden-Rynsch 2004: 116, [...] suppl. K. B.).

[30] "Das Emporkommen der mächtigen Mätressen und der eleganten Kurtisanen prägte ebenso wie die Umgangsformen und die Entscheidungen ihrer Kavaliere auch die Geschmacksbildung und das Selbstbewußtsein der Frau von Stande, die sich früher entsetzt von diesen berüchtigten Damen abgewendet hatte" (Heyden-Rynsch 2004: 116).

present her, the ballerina, in the best possible way[31]. The male dancer is just a "porteur," the "third leg" of the ballerina so to speak which allows the female principle to grow up to a "nearly absolute domination" (Zacharias 1962: 58.). Only much later, in the early 20th century, does the male dancer regain a position of artistic equality with the ballerina. Perhaps the most famous example of this is found in the ballet russe – the Russian ballet – where Waslaw Nijinsky provides a male artistic counterbalance to the dominant position of the ballerina.

Klein (1992) interprets the role of the ballerina as a demonstration of the relation of dependence between the sexes to the disadvantage of the woman: "Even as a hovering ideal the female dancer was a function of the man; he helped her on the stage in order to stage-manage her as a picture[32]" (ibid.: 126, transl. K. B.). She becomes an 'active object of suppression' by projecting the male desires as well as his repressions on the female dancer, and particularly on her body (ibid.: 93-129). In this statement, Klein fails to consider that male and female dancers serve two completely different functions which are both indispensable for the preservation of a hierarchically structured social order. Thus, the characterization of the relationship could be one of interdependence.

The female ballet dancer of the Romantic era acts indeed as a person with two natures (ibid.: 117) with a stage identity and a social identity. On the stage she can represent Nordic divinities, charm spirits as well as fairies, elves and witches risen from the world of fairytale. All of them have in common – independently whether they are positively or negatively proscribed – that they are not social beings but creatures with supernatural power.

In "La Sylphide" (1832) it is a lovely, winged forest fairy or sylph, whom James falls in love with. The witch Madge tosses him a magic scarf and instructs him to wind the scarf around the sylph's shoulders to bind the sylph to himself so she cannot fly away. But the moment the sylph is captured her wings fall off and she dies in his arms (Pasi 1980: 95 f.; Zacharias 1962: Figure 25).

In "Giselle" (1841), the beautiful peasant girl (after whom the ballet is named) falls in love with Duke Albrecht, a nobleman disguised as a peasant. Although he is engaged to a princess, he flirts with Giselle. After she learns the truth about her beloved one, she dies and turns into a female spirit, a Wili. Normally, Wilis were jilted women who would rise after their deaths to seek revenge upon men by dancing them to death. In this case, Giselle becomes a protector, when Myrtha, the Queen of Wilis, sentences Duke Albrecht to death. Giselle protects him from Myrtha and other Wilis, because her love for Albrecht is still undiminished (Pasi 1980: 102 f.).

In the opinion of Klein (1992) middle-class male principles reduce women to "an unfortunate loving maid or a threatening witch; dance becomes therefore an appropriate form of body movement, incarnating as a female dancer the Christian principle of sin[33]" (ibid.: 115, transl. K. B.). The morality of the middle-class and of Christianity meet one another to

[31] With the appearance of the Russian or Tsarist ballet, the emancipation of man virtually takes place: Instead of the passive side-roles, the male dancer claim autonomous, active dance parts for themselves. Serge Diaghilew, for example, is an important feature in this ‚renaissance' of male ballet dance of the last century (Zacharias 1962: 61).

[32] "Noch als schwebendes Ideal stand die Tänzerin in Abhängigkeit von dem Mann; er half auf der Bühne, sie als Bild zu inszenieren" (Klein 1992: 126).

[33] "...in ein unglücklich liebendes Mädchen und eine bedrohliche Hexe, der Tanz als die ihr angemessene Bewegungsform, die Tänzerin als Inkarnation des christlichen Sündenprinzips" (Klein 1992: 115).

"imposing on the women a voluntary or forced life's work and the aim of renunciation[34]" (ibid.: 117).

This argumentation ignores the fact that the ritual function of the ballerina places into another context of consciousness – social and the interpersonal relations. Romantic ballet unfolds on two different levels which complete each other; the first is the social real world and the second is the supernatural one, "the magic-symbolic nightly myth[35]" (Pasi 1980: 104, transl. K. B.). The female dancer undertakes the task of a mediator in this sense. She is the muse of dance par excellence, the divine soul, who mediates between the transcendental world of enlightenment and the sensuous sphere of mundaneness.

Here, you find to different dancing personalities. The Indian dancer – as being understood in Indian advaita philosophy to be a priori the owner of the ātman[36], of inner completeness – is able to discover and bring out her self-realising existence from within (like an "instatic" process), while the ballet dancer must undergo a transformation or leave her body in order to attain access to the "higher" state of being situated beyond her bodily existence (like an "ecstatic" process). From this point of view the female ballet dancer is much like a shaman[37] in his function as a mediator when he leaves his body and externalises himself to gain ultimate knowledge from the world beyond matter. The female ballet dancer actually becomes like someone "possessed" by the higher reality in order to pass on the enlightening wisdom to mankind.

Through this metamorphosis the ballerina becomes indeed (and necessarily) an "unreal," an absolutely spiritual being, because only with this new, supernatural identity she is able to manoeuvre herself (and together with her all spectators) into a "true reality," i. e. in Plato's words the sphere of the Beauty or the Good.

Being neither human nor woman, she overcomes her "natural" (i. e. biological) appearance, which explains also the denial of her femaleness as well as her earthboundness, e.g., the denial of her pelvis (Klein 1992: 127), the narrow wasp waist (Liechtenhan 1983: 76) or the emphasis on weightless dance movements. The ancient Greek idea of beauty as a being-in-itself and therefore the supreme and dominant principle of the Whole, turns into the ideal of an "asexual beauty" (Sorell 1985: 244). Here, at this point, a strong structural approach to the Christian ideal of Virgin Mary takes place, because both women are disembodied to the uttermost extend and thus "asexual" (Klein 1992: 120 ff.).

Holy Mary as well as the ballerina represent a high value of the Christian occident: "purity" expresses above all an aesthetic and moral meaning. The Virgin Mary, the pure handmaid of the Lord and the noble woman, to whom the minnelied was dedicated, both are ideologically embedded in society as the "epitome of female eroticism and beauty" (Klein 1992: 60). The ballet dancer unites both identities on the stage: the "purity of her presence" as well as the "beauty of the mind". She becomes the female idol of the Romantic era, which "really exists but nonetheless remains distant from everyday reality in order to maintain the dream of the spiritual-pure one and of the unattainable ideal[38]" (Sorell 1985: 229, transl. K. B.). Longing for the never attainable absolute as well as an awareness of this utopian dream

---

[34] "...über eine freiwillige oder erzwungene Lebensaufgabe und Entsagung der Frauen (Klein 1992: 117).

[35] "...des magisch-symbolischen nächtlichen Mythos" (Pasi 1980: 104).

[36] See Barboza (in prep.) "Dancing realization".

[37] For Zacharias (1962) the jumping movement symbolizes the flight of the shaman. For further details see Barboza (in prep.) "Concept of earth in dance" and "Dancing realization".

[38] "...wirklich ist und doch von der alltäglichen Realität entfernt bleibt, um den Traum des Spirituell-Reinen und des unerreichbaren Ideals aufrechtzuerhalten" (Sorell 1985: 229).

as a life-long condition exists as a leitmotiv not only in Romantic ballet but also in Romantic literature.

Perhaps the Romantic concept of purity explains the change of costume styles in ballet. Instead of the colourful court costumes of the noble dancers (during the Renaissance and Baroque), white tulle, gauze and brocade are worn; from the multicoloured court ballet the Romantic "ballet blanc[39]" emerges (Balcar 1957: 38; Zacharias 1962: Figure 9).

Against this background the female dancer enjoys during performances a high artistic and – as long as she holds the function of a mediator – a high temporary ritual status. This, however, is not identical to her social status in daily life.

## The Female Ballet Dancer in Romantic Period Social Life

In social life the female ballet dancer – mostly originating from families of lower socio-economic status or from intergenerational dancer's families – experiences no social acknowledgement. Peter the Great, whose love for ballet kindled a real contest around the promotion of young ballerinas, recruited most dancers from the "corps" of the daughters of his slaves. Very talented dancers were sold or donated as a gift to the Royal Theatre, e.g., the first prima ballerina Pauline Kovalevsky (Sorell 1985: 202).

Due to low wages, most of the dancers scarcely had enough to live on and needed to prostitute themselves to make a living (Klein 1992: 117 f.). The expression "ballet rat[40]" uncovers the low regard for female dancers and is used by "cooperative" theatre managers whenever they made available one of their ballerinas for "private occasions" to a rich person (ibid.: 118). Klein (1992) characterizes the situation of the woman as a double sale of her own body, viz. on the stage and in social life (ibid.: 118). Their social existence is thus negated doubly, in the theatre as well as in everyday life. Either "saint" or "whore," as personified beauty or as societal outlaw, a female dancer moved at the opposite poles of a society without ever having a proper place in that society (Klein 1992: 120). Under this compound condition the female ballerina becomes a tragic figure, born out of the rise of Romantic ballet.

In contrast, the male dancer remains the "only human constancy" in ballet (Klein 1992: 116). The roles which male dancers present on stage, are consistent with their social status. To be more precise, the male ballet dancer of the Romantic period likewise possesses two natures. On the stage he is the passive "porteur" who dances in the shadow of the ballerina, which means he takes only indirectly part in the active transformation process. But, in the social reality of the middle class the man is a powerful, mighty and rational person, who, by virtue of his maleness, possesses a higher social rank than the ballerina. Thus, he can also be seen as a figure that subordinates the woman.

If we regard both gender roles from a ritualistic point of view, each sex demonstrates on the stage an "anti-structure" (Turner 1989) which opposes respective social norms or gender identities ("structure")[41]. Both of these, the transforming as well as the preserving, are indispensable for the continuance of a society. The problem with this is not the different distribution of roles. Rather, it is the shift of social context; classical ballet, originally an

---

39 "Ballet blanc" means literally 'white ballet'.

40 During the 19th century emerged the nickname ‚le Rat' (the rat) for dance pupils and young female members of the ballet corps (Liechtenhan 1983: 76).

41 For further details see Barboza (in prep.) "Dancing action".

aristocratic art, is conceived so that it ritually protects, or rather renews, a gender-based and hierarchical societal structure.

After having outlined the characteristics of gender roles in their prevailing socio-historical context, the following discussion considers the thesis of Zacharias (1962). Zacharias identifies the ballet dancer as having a dual sexuality. With this context in mind, the representative meaning of the "pas de deux," the duet, in ballet is discussed.

## The Dancer as Hermaphrodite?

According to Zacharias (1962), the classical ballet dancer overcomes "the painful discord, caused by the dualism of the gender, because he embodies the human-androgynous unit[42]" (ibid.: 52, transl. K. B.). As a hermaphrodite being aware of its androgynous origin, the dancer overcomes the antagonism of gender in order to attain, in a humanistic sense, his entireness: "The dance ... reaches a depth ..., where the primal substance of the soul does not differentiate any longer between male and female, but is only defined as being 'human'[43]" (Terpis, quoted in: Zacharias 1962: 51, transl. K. B.). Similarly, Lorenz (1987) assigns the "androgynous[44]" dancer to an aura of "sexual inviolability" and of "no erotic desire" (ibid.: 116 f.).

In Indian dance, there is a conception of an androgynous primordial being, as well. Especially in South India, it is the adored deity śiva as ardhanārīśvara (half female God) or as naṭarāja (king of the dance) who embodies and unites both gender categories in one form. In this, we find an ancient Greek correspondence in Plato's (1991) idea of the "manwoman" which he refers to in his "Symposium." Here, Aristophanes, a great poet of Athens, participates in a symposium and explains in his speech why one feels "whole" when you have found your love partner. According to classic mythology, the state of Man was not what it is now. There were three sexes, i.e., the male, the female and a third having both sexes together (ibid.: 99). The shape of this hermaphrodite was quite round, they had four arms and four legs and when running fast they rolled over and over like a tumbler. Since they were strong and had great ambitions, they attacked the gods. Finally, Zeus sliced each of them in two halves, in a female and a male one. Ever since that time, men and women have searched for their lost counterpart (ibid.: 103). This desire for the whole and the pursuit of it is called "Love" (ibid. 107).

In Christian and Jewish written records (in the Bible and the Torah) of the creation of Man according to second account of man's creation in the Book of Genesis, the ancient Greek idea of the hermaphrodite seems to disappear: Adam, the first being of mankind in a generic sense ("Man"), is at the beginning without gender, because there is nothing to compare him/her/it "against." He even encompasses[45] his opposite, Eve, the woman. It is only with the

---

[42] "...die qualvolle Zwietracht, die mit dem Dualismus des Geschlechts gegeben ist, indem die menschlich-androgyne Einheit in ihm eine Verkörperung findet" (Zacharias 1962: 52).

[43] "Der Tanz ... reicht in eine Tiefe ..., wo die seelische Ursubstanz nicht mehr unterscheidet zwischen männlich und weiblich, sondern nur noch als >menschlich< bezeichnet werden kann" (Terpis, quoted in: Zacharias 1962: 51).

[44] I hold the view that the term "androgynous" the way Lorenz (1987) uses it, is to observed critically, because she describes with this term more the masculinization of the female body ("slim, little-bosomed, distinctively trained leg muscles, boyish" ibid.: 116) than a hermaphrodite being.

[45] Dumont (1980) calls it the "the encompassing of the contrary" (ibid.: 239).

creation of Eve from his ribs, that Adam manifests himself as "male." Thus, in Western cultures man represents both, the "whole" ("Man") as well as the individual ("man"), whereas a woman is part of the opposition pair "man-woman" (cf. Dumont 1980).

This Christian influence is surely one reason why the idea of hermaphrodite does not represent a socially recognised category in Western societies, while the deity śiva as a hermaphrodite is alive all over India, especially in the south of India. An additional factor might be attributed to the appearance of natural sciences in the West. During the Age of Enlightenment the classification of socio-cultural gender takes place exclusively according to a biological binary i.e., male and female.

Current debate about transsexuality[46] supplies a contemporary example about how a male or female might transition to the contrary gender, yet still remain, in the minds of some, in their original socio-cultural category as either "man" or "woman". The option to live both sexes at the same time or define oneself as "neutra" is not widely accepted notwithstanding that some languages, e.g., German, have a grammatical neuter and thus, the possibility of social or linguistic expressions beyond the traditional binaries exists.

Biomechanical analysis of ballet does not reveal gestures or steps that can be identified as androgynous. It may be true that the male dancer must not be or appear as too male, the ballerina as not too female[47]" (Zacharias 1962: 51, transl. K. B.) however, there is a clear assignment of male and female roles.

That means that each dancer represents *one* particular gender category which is defined as either male or female. This contrasts with the Indian dance in which the gender roles might switch from one second to the next, and as often as necessary. In a certain way this interrelated play of the roles results from a necessity, because the Indian dance bharatanāṭyam is primarily a solo dance, while the majority of French Ballet choreographies involves groups of two or more. Even the pas de deux, the core of ballet dance – on which I will discuss separately – must be danced by at least two people, as the French expression (which literally means "dance for two") already indicates.

The classification system of the ballet arranges likewise the movement patterns according to the gender. Thus, for example, the step sequence of the pas de bourrée or the battement of the male dancer looks different from the female one[48]. In many choreographies of 17th and early 18th-century court and social dances they are mirror images. In theatrical traditions, solo dances can have very different steps.

Blasis (1828) emphasizes expressively the gender-depending movement patterns. "Men must dance in a manner very different from women; the temps de vigueur, and bold majestic execution of the former, would have a disagreeable effect in the latter, who must shine and delight by lithesome and graceful motions, by neat and pretty terre-à-terre steps, and by a decent voluptuousness and abandon in all their attitudes" (ibid.: 94 f.).

Additionally, there are dance combinations which are destined for one particular gender:

a) To the domain of the man belong generally the athletic leaps, especially the turns in the air[49] (Zacharias 1962: 113) as well as the jumps with crossed legs[50] (ibid.: 102).

---

46 For further references see Hirschauer (1993; 1994; 1996) and Lindemann (1993; 1994).

47 "...ein Ballerino nicht extrem männlich, eine Ballerina nicht extrem weiblich sein bzw. wirken darf" (Zacharias 1962: 51).

48 See the biomechanical data.

49 "Tour en l'air".

Above all, these are formations who detach themselves of the earth in order to symbolize "spirit" and "strength" (ibid.: 91). In general, these virtuous powerful movement items are classified as "allegro dance" (Balcar 1957: 77; Niehaus 1954: 19)

b) The characteristics of the Romantic female dance are the typical elegant arm movements[51], the dance on points and special turns like the tours chaînés (Zacharias 1962: 113). They reflect the "botanical" nature of the woman which opens and closes at times (ibid.: 91). These rhythmical slow movements are labelled as "adagio[52]-dance" and culminate in the classical "pas de deux" (Balcar 1957: 76 f., Niehaus 1954: 87)

We find the final connection of the adagio and allegro dance in the Russian Ballet at the end of the 19th century, "which technique, fixed in a comprehensive system, is until today still the basis of the classical dance form[53]" (Balcar 1957: 77, transl. K. B.).

Zacharias (1962) differentiates both the dance of men women on the basis of uniting two contrasting features, which are different in each case:

1) The dance of men unites the two primeval opposites of the Christian Spirit ("solar manliness") and of the heathen life of instincts ("faunic-chthonic manliness") to an "entireness of man" (Zacharias 1962: 53 f.).

These two aspects are represented par excellence by the two well-known male ballet dancers, viz. Louis XIV of France, known as the "Sun King" who defeats the darkness, and the Russian dance talent Nijinsky as an animal creature of instincts: "Louis XIV in his role of the Sun and Nijinsky as a faun embody the two basic possibilities of male existence which result from the polarity of spirit and instinct[54]" (ibid.: 54, transl. K. B.; see also Figure 4, 5).

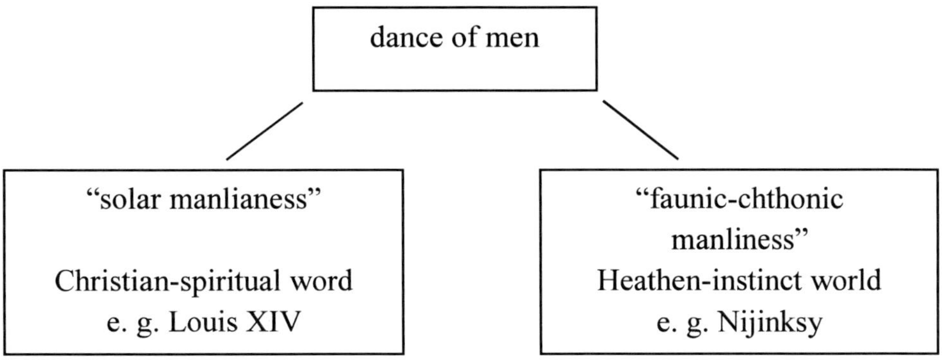


Figure 41. Model of the dance of men (Barboza).

50 "Entrechat".

51 "Port de bras".

52 Besides its meaning in music (slow, expressive) the term "adagio" (ital.) or "adage" (fr.) has its own meaning in dance which describes the ease, the turns and the linear elegance of the (female) dancer (Regner 1964: 417).

53 "...dessen Technik, in einem umfassenden System fixiert, auch heute noch die Grundlage des klassischen Tanzes bildet" (Balcar 1957: 77).

54 "Ludwig XIV. in der Sonnenrolle und Nijinsky als Faun verkörpern die beiden Grundmöglichkeiten männlicher Existenz, die sich aus der Polarität von Geist und Trieb ergeben" (Zacharias 1962: 54).

Even if male classified dance shows mental as well as instinct driven aspects of our existence, it remains the fact that solar manliness subordinates the faunic-chthonic one. Reactions or rather vehement protests of some dance critics give us blunt information about this. Regarding Nijinsky's debut: "One saw only an unchaste, common faun with animal-erotic and shameless gestures. That was everything[55]" (Zacharias 1962: 54, transl. K. B.).

2) Zacharias (1962) recognizes a similar dualism in the dance of women. The prima ballerina Taglioni and her "spiritual hovering" dance symbolizes the "bright" aspect of the "celestial world soul (anima mundi)" while Elßler, full of "earthy naturalness" and "vivid sensuality", represents the dark, earthy aspect of existence (Zacharias 1962: 58 and Figure 6; Sorell 1985: 243).

Barchan (1913) reinforces this conception of dance stereotypes. In his chapter *Die Arten* (1913) he assigns the "danse d'élévation" to the elegance and grace appearing "ballon[56]-dancer", who is able to "transform the dance into spirituality through intelligence, soul and ideational demands[57]" (ibid.: 9, transl. K. B.). Her spirituality makes the dance actually to a "heavenly" art form, while the virtuous dance belongs to the realm of the "danse terre à terre". Latter one spreads joy, earthiness and so on, but not from a mental disposition but because the "blood" of the dancer determines the way of dancing (ibid.: 9). And, the "earthy-sweet" Camargo and the "refined" Sallé (ibid.: 9 f.) are two typical French representatives of these two "dance-characters."

As late as Balcar (1957) the old dualism between narrow and far moving dance is observed in the "danse d'élévation" and in the "danse terre à terre" (ibid.: 31).

Also in the dance of women a separation takes place according to common Christian conception: Taglioni is charac-terised as a "Christian," Elßler as a "Heathen" dancer (Zacharias 1962: 60). Both characters united, the "Great-Female" or rather the "world soul" becomes present on the stage:

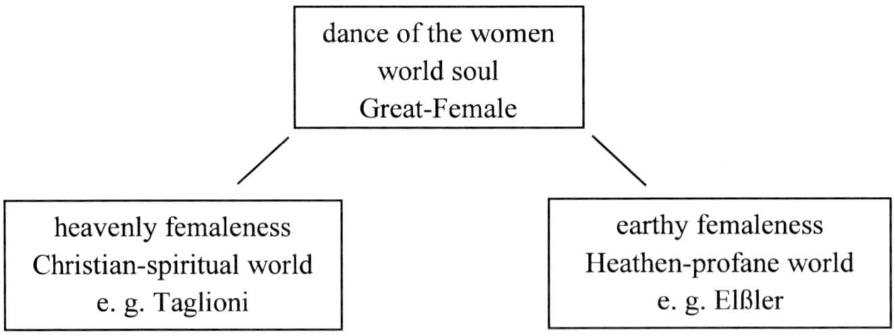


Figure 42. Model of the dance of women (Barboza).

---

[55] "Wir haben nur einen unkeuschen, gemeinen Faun gesehen mit tierisch-erotischen und schamlosen Gesten. Das war alles" (Zacharias 1962: 54).

[56] The ballet term "ballon" refers to the ability of a dancer to appear to hold a position in the air to create the illusion to glide across the floor and through the air as if defying the law of gravity.

[57] "...durch Geist, Seele, ideelle Forderungen, den Tanz in Geistigkeit umzusetzen" (Barchan 1913: 9).

If we compare both types of dancers with one another, then it is noticeable that *both* men and women reflect a dualistic existence which unite different aspects of reality. Here, in the world of dance, the conception of the poet and painter Victor Hugo becomes true, who demands in 1827 for any artistic expression the "freedom of the creative spirit": "Not all creation is human and beautiful, ugliness exists next to beauty ... art needs to do what nature does, mixing and forming of opposites ... the body and the soul, the beast and the intellect ... everything is connected ...[58]" (ibid., quoted in: Sorell 1985: 238, transl. K. B.).

## The Pas De Deux

A special role[59] is granted to the ballerina as a representative of the "world soul" who mediates[60] between the celestial and profane reality to animate the living. This is expressed beautifully in the pas de deux. The pas de deux (dance for two) is a duet which traditionally consists of five parts: entrée[61], adagio[62], variation ballerino/solo (pas seul) of the male dancer, variation ballerina/solo (pas seul) of the female dancer, coda[63] (Zacharias 1962: 64). During the entrée and adagio both the male and female dancers appear together, for the soli they split up, only to reunite in the final coda (ibid.: 64 f.).

In the classical pas de deux the ballerina is without doubt the principal figure. She is the one who is carried and presented by the male dancer, while he supports her for the intricate poses, lifts her, holds her during the pirouettes, and majestically struts around her (Zacharias 1962: 65). During the promenades (en promenade) the male dancer walks proudly around the female dancer to pay honour to her. As in the case of walking around a shrine, a temple or holy place, the male dancer absorbs – in place of the audience – the magic, transforming energy of the ballerina (ibid.: 65 f.).

The many lifting figures indicate a clear gender separation: The man is the carrying one, the woman the person to be carried, the same way like the body carries the soul, since "the dancer as the porteur symbolizes the body, carrying the soul; the with his body action lifted ballerina reflects the soul, transforming him[64]" (Zacharias 1962: 66, transl. K. B.). In addition to the lifting figures, the pirouettes – as are the single or multiple rotations of a dancer called – Zacharias (1962) interprets as the unification of the male and female principle, because the ballerina encloses with her hand the central or index finger of the male dancer (Zacharias 1962: 66).

On the basis of this analysis it becomes obvious that the ballerino and the ballerina realize two fundamentally different functions. Both aspects are equal and can mutually complete each other during the dance. In this context one can speak about an interdependent

[58] "Nicht alles in der Schöpfung ist menschlich und schön, das Häßliche existiert neben dem Schönen ... Kunst muß das tun, was Natur tut, Gegensätzliches mischen und gestalten ... den Körper und die Seele, die Bestie und den Intellekt ... Alle Dinge sind miteinander verbunden ... " (Hugo, quoted in: Sorell 1985: 238).

[59] The mediator role of the female dancer will be taken up again in Barboza (in prep.) "Dancing realization".

[60] The meaning of the soul in the antique Greek context will be elaborated in Barboza (in prep.) "Dancing expression/aesthetics".

[61] The French term "entrée" means literally "entrance" and describes the beginning of a dance part or dance section (Regner 1964: 426).

[62] Here, the female and male dancer still work together, but the emphasis is put on the presentation of the female aspect (Regner 1964: 417).

[63] The Italian word "coda" means "tail" and describes the last part of a choreography (Regner 1964: 422).

[64] "...der Tänzer als Porteur symbolisiert den die Seele tragenden Körper, die von ihm gehobene Tänzerin die vom Körper getragene und ihn wandelnde Seele" (Zacharias 1962: 66).

relationship. Related to the whole, there is neither any inferiority or superiority, but equality, because both dancing gender identities need each other to be complete.

Due to the clear division of gender roles into two parts, viz. the male and the female, one cannot speak about an androgynous dancer but only about a binary gender dance structure. The reason for it, I presume, is the idea that through the perfect interaction of man and woman the boundaries of gender identity as a single, isolated identity are almost resolved – the perfect state (of being one) will never be achieved completely. But the knowledge about this unattainable aim is undoubtedly the leitmotiv of the Romantic era as such.

## Summary of the Comparison of the Gender Concept

1) The gender roles of ballet dance are – depending on the epoch –subject to strong changes:
    a) During the Renaissance noble women dance partially the leading role, but the idealised female one is finally subordinated to the man (king).
    b) The Baroque court ballet (1600-1680) is predominantly danced by men who take over also female roles. Similarly to Indian dance the French ballet becomes at this historic period a "gender performing act," enabling to cross the biological ascription. The difference between both dance styles consists of the fact that in the Western culture only the man is enabled to change the gender role, while it is open to both genders in India. Besides this, the male role performance of the Baroque is based on the social sanction of women, whereas it is in India a divine (and thus, generally accepted) concept.
    c) In the Rococo ballet an emancipation of the woman takes place on the stage.
    d) The Romantic ballet is characterized by the supremacy of the ballerina. Female and male dancers demonstrate their gender roles analogous to their biological sex.
2) The exceptional feature of the woman is situated in her function of a mediator which enables her to mediate between the profane and transcendental world. In Indian dance this function is independent of the gender, but valid for all dancers.
3) The nature of the ballet dancer has two aspects: On the stage the woman possesses a status-high, ritual stage identity and a status-low social identity. The man is only passively involved in the staging, however, in the social reality of the middle class he points the way. Both gender roles demonstrate thus a ritual "anti-structure" (Turner 1989), contrasting each one the respective social structure and its norms.
4) We find the temporarily high ritual status and the low social status also among traditional Indian artists (craftsmen, dancers, musicians).
5) The ballet dancer is not, in contrast to Indian dancers, a hermaphrodite. Also the biomechanical data of the ballet dance movements reflect a clear gender specificity, whereas Indian dancers do not.
6) The dance of women as well as of men in Romantic Ballet reflects a moral-dualistic nature, seen in the dualistic pairs of opposites, viz. of heaven and earth or Christian and Heathen values. Accordingly, in the social acceptance the "celestial" aspect subordinates the "earthen." In Indian dance the earthen and the celestial attributes are equal, both necessarily complete each other's existence and hence build an interdependent pair of oppositions.

7) In ballet both genders are presented separately, in order to unite in harmonious interaction. The perfect harmony resolves almost the boundaries of gender identity, but they are never perfectly overcome – in the Romantic epoch the knowledge about this unattainable aim is the basic leitmotiv as such. In contrast to this, Indian dancers embody permanently a divine androgynous existence which is resolved finally in a spiritual Neutra, the absolute consciousness which is without any distinction.

## REFERENCES


Balcar, J. A. (1957). *Das Ballett: Eine kleine Kulturgeschichte* [*The Ballet: A small cultural history*]. München: Winkler.

Ballard, R. (1653). *Ballet Royal De la Nuit* [*Royal Ballet of the Night*]. http://www.google.de/url?sa=t&rct=j&q=&esrc=s&source=web&cd=9&ved=0ahUKEwjm0PTVpv7RAhUGiCwKHc59AY4QFghZMAg&url=http%3A%2F%2Fnephilimrevelation.free.fr%2FIMG%2Fpdf%2FN0072470_PDF_1_76.pdf&usg=AFQjCNGhHzkPV80KfqPXvr6Mu_aBeiAP1w&sig2=LLXSQivhDAF2abwxsQBTDQ&bvm=bv.146094739,d.bGg

Barchan, P. (1913). *Die Arten* [*The species*]. In O. Bie (Ed.), Anna Pawlowa (pp. 9-21). Berlin: Bruno Cassirer.

Bharata (n.y.). The Nāṭya Śāstra of Bharatamuni. *Rāga Nṛtya Series No. 2*, Delhi: Sri Satguru Publications.

Blasis, C. (1828). *The code of Terpsichore: A practical and historical treatise, on The Ballet, Dancing and Pantomime. With a complete theory of the Art of Dancing. Intended as well for the instruction of amateurs as the use of professional persons.* London: James Bulcock, (NP 1976, Brooklyn: Dance Horizons).

Blomkvist, M. (1996). François de Lauze und seine Apologie de la danse (1623). In S. Dahms, & S. Schroedter (Eds.), *Tanz und Bewegung in der Barocken Oper, Kongressbericht Salzburg* 1994 [François de Lauze and his apology de la danse (1623). In S. Dahms, & S. Schroedter (Eds.), *Dance and Movement in the Baroque Opera, Congress Report Salzburg* 1994] (pp. 31-43). Innsbruck, Wien: Studien-Verlag.

Blumenberg, H. (1996). *Arbeit am Mythos* [*Work on the myth*]. Frankfurt a. M.: Suhrkamp.

Burckhardt, J. (1976). *Die Kultur der Renaissance in Italien: Ein Versuch* [*The culture of the Renaissance in Italy: an attempt*]. Stuttgart: Kröner.

Butler, J. (1991). *Das Unbehagen der Geschlechter* [*The discomfort of the sexes*]. Frankfurt a. M.: Suhrkamp.

Cassirer, E. (1990). *Versuch über den Menschen: Einführung in eine Philosophie der Kultur* [*Trial on Man: Introduction to a Philosophy of Culture*]. Frankfurt a. M.: Fischer.

Coomaraswamy, A. K. (1991). *The dance of Shiva: Fourteen Indian Essays.* New Delhi: Munshiram Manoharlal Publishers.

Dumont, L. (1980). *Homo hierarchicus: The caste system and its implications.* Chicago/London: The University of Chicago Press.

Eliade, M. (1985). *Yoga: Unsterblichkeit und Freiheit* [*Yoga: immortality and freedom*]. Frankfurt a. M.: Suhrkamp.

Feuillet, M. (1700/1701). *Choregraphie ou l'art de de'crire la dance, par caracteres, figures et signes de'monstratifs, Avec lesquels on apprend facilement de soy-même toutes sortes de Dances. Ouvrage tres-utile aux Maîtres à Dancer & à toutes les personnes qui s'appliquent à la Dance* [*Choreography or the art of describing dance, by characters, figures and signs of demonstrations, with which one can easily learn from all kinds of Dances. A very useful book for Dancer Masters and all those who apply to Dance.*]. (Faksimile NP 1981, University Microfilms International, Michigan/London: Ann Arbor).

Fichte, J. G. (1971). Grundlage des Naturrechts nach Principien der Wissenschaftslehre. In I. H. Fichte (Ed.), *Fichtes Werke III. Zur Rechts- und Sittenlehre I* [Basis of natural law according to principles of the science of science. In I. H. Fichte (ed.), *Fichte works III. To the law and ethics I*] (pp. 1-385). Berlin: de Gruyter & Co.

Günther, H. & Schäfer, H. (Eds.) (1959). *Vom Schamanentanz zur Rumba: Die Geschichte des Gesellschaftstanzes* [*From Shaman Dance to Rumba: The History of Ballroom Dancing*]. Stuttgart: Ifland.

Heyden-Rynsch, V. van der (2004). *Das Spiel der Verführung: Liebe und Galanterie im Wandel der Zeiten.* Darmstadt: Wissenschaftliche Buchgesellschaft [*The game of seduction: love and gallantry through the ages.* Darmstadt: Scientific book company].

Hirschauer, S. (1996). Wie sind Frauen, wie sind Männer? Zweigeschlechtlichkeit als Wissenssystem. In C. Eifert, A. Epple, & M. Kessel (Eds.), *Was sind Frauen? Was sind Männer? Geschlechterkonstruktion im historischen Wandel* [How are women, how are men? Two-genderedness as a knowledge system. In C. Eifert, A. Epple, & M. Kessel (Eds.), *What Are Women? What are men? Gender construction in historical change*] (pp. 240-256). Frankfurt a. M.: Suhrkamp.

Hirschauer, S. (1994). *Die soziale Fortpflanzung der Zweigeschlechtlichkeit. Kölner Zeitschrift für Soziologie und Sozialpsychologie* [*The social reproduction of the two sexes. Cologne Journal of Sociology and Social Psychology*], 46, pp. 668-692.

Hirschauer, S. (1993). *Die soziale Konstruktion der Transsexualität: Über die Medizin und den Geschlechterwechsel* [*The Social Construction of Transsexuality: About Medicine and Gender Change*]. Frankfurt a. M.: Suhrkamp.

HKW (Haus der Kulturen der Welt)/ Staatliche Museen zu Berlin (Ed.) (1992). *Palast der Götter: 1500 Jahre Kunst aus Indien* [*Palace of the Gods: 1500 years of art from India*]. Berlin: Reimer.

Jung, V. (2001). *Körperlust und Disziplin: Studien zur Fest- und Tanzkultur im 16. Und 17. Jahrhundert* [*Body Lust and Discipline: Studies on festive and dance culture in the 16th and 17th centuries*]. Köln/Weimar/Wien: Böhlau.

Klein, G. (1992). *Frauen Körper Tanz: Eine Zivilisationsgeschichte des Tanzes* [*Women's Body Dance: A Civilization History of Dance*]. Weinheim/Berlin: Quadriga.

Lévi-Strauss, C. (1993). *Die elementaren Strukturen der Verwandtschaft* [*The elementary structures of kinship*]. Frankfurt a. M.: Suhrkamp.

Liechtenhan, R. (1983). *Vom Tanz zum Ballett: Eine illustrierte Geschichte des Tanzens von den Anfängen bis zur Gegenwart* [*From dance to ballet: An illustrated story of dancing from the beginning to the present*]. Stuttgart/Zürich: Belser.

Lindemann, G. (1994). Die Konstruktion der Wirklichkeit und die Wirklichkeit der Konstruktion. In T. Wobbe & G. Lindemann (Eds.). *Denkachsen. Zur theoretischen und institutionellen Rede vom Geschlecht* [The construction of reality and the reality of construction. In T. Wobbe & G. Lindemann (Eds.). *Thinking axes. For theoretical and institutional talk about gender*] (pp. 115-141). Frankfurt a. M.: Suhrkamp.

Lindemann, G. (1993). *Das paradoxe Geschlecht. Transsexualität im Spannungsfeld von Körper, Leib und Gefühl* [*The paradox gender. Transsexuality in the field of tension of body, body and feeling*]. Frankfurt a. M.: Fischer.

Lorenz, V. (1987). *Prima Ballerina: Der zerbrechliche Traum auf Spitzen* [*Prima Ballerina: The fragile dream on tips*]. Frankfurt a. M.: Athenäum.

Niehaus, M. (1954). *Ballett*. München: Prestel.

Noverre, J.-G. (1760). *Lettres sur la danse et sur les ballets* [*Letters on dance and ballets*]. Lyon: Chez aimé Delaroche.

Pasi, M. (Ed.) (1980). *Ballett: Eine illustrierte Darstellung des Tanztheaters von 1581 bis zur Gegenwart* [*Ballet: An illustrated depiction of the dance theater from 1581 to the present*]. Wiesbaden: Drei Lilien.

Platon (1991). *Platon: Sämtliche Werke in zehn Bänden. Griechisch und Deutsch* [*Plato: All works in ten volumes. Greek and German*]. Bd. IV. Hippias minor. Symposion. Phaidon. K. Hülser (Ed.). Frankfurt/Leipzig: Insel.

Prest, J. (2006). Female Roles in Court Ballet I: Men Playing Women. In (ibid.) *Theatre Under Louis XIV. Cross-Casting and the Performance of Gender in Drama, Ballet and Opera* (pp. 77-101). London: Palgrave Macmillan.

Ranke-Graves, R. (1989). *Griechische Mythologie: Quellen und Deutung* [*Greek mythology: sources and interpretation.*]. Reinbek/Hamburg: Rowohlt.

Regner, O. F. (1964). *Reclams Ballettführer* [*Reclams ballet leader*]. Stuttgart: Reclam.

Sadasivan, K. (1993). *Devadasi System in Medieval Tamil Nadu*. Trivandrum: CBH Publications.

Sarabhai, M. (1992). *Understanding Bharata Natyam*. The Darpana Academy of Performing Arts, Ahmedabad.

Sator, K. (2003). Tanz und Homosexualität. Sexuelle Identitäten hinter und auf der Bühne. In Fachverband Homosexualität und Geschichte e.V. (Ed.): *Invertito - Jahrbuch für die Geschichte der Homosexualitäten - 5. Jahrgang - Zwischen Camouflage und Provokation: Homosexualitäten in Kunst und Kultur* [Dance and homosexuality. Sexual identities behind and on stage. In Fachverband Homosexualität und Geschichte e.V. (Ed.): *Invertito - Yearbook for the History of Homosexuality - 5th Year - Between Camouflage and Provocation: Homosexuality in Art and Culture*] (pp. 67-100). Hamburg: Männerschwarm Verlag.

Sorell, W. (1985). *Der Tanz als Spiegel der Zeit: Eine Kulturgeschichte des Tanzes. Wilhelmshaven* [*Dance as a Mirror of Time: A Cultural History of Dance. Wilhelmshaven*]: Heinrichshofen.

Stutley, M. & J. (1977). *A Dictionary of Hinduism: It's Mythology, Folklore and Development 1500 B. C. – A. D. 1500*. London: Routledge and Kegan Paul.

Turner, V. W. (1989). *Das Ritual: Struktur und Anti-Struktur* [*The ritual: structure and anti-structure*]. Frankfurt a. M./New York: Campus.

Vatsyayan, K. (1977). *Classical Indian Dance in Literature and the Arts*. Sangeet Natak Akademi, New Delhi.

Vosz, R. (1869). *Der Tanz und seine Geschichte. Eine kulturhistorisch=choreographische Studie: Mit einem Lexikon der Tänze* [*The dance and its history. A cultural-historical = choreographic study: With a dictionary of dances*]. Berlin: Seehagen.

Zacharias, G. (1962). *Ballett – Gestalt und Wesen: Die Symbolsprache im europäischen Schautanz der Neuzeit* [*Ballet - shape and essence: the symbolic language in the European visual dance of modern times*]. Schauberg/Köln: DuMont.

Zimmer, H. (1974). *Myths and symbols in Indian art and civilization*. Bol-lingen Series VI, Princeton: Princeton University Press.

In: Advances in Arts Biomechanics
Editors: Gongbing Shan and Peter Visentin
ISBN: 978-1-53615-641-6


*Chapter 8*

# EFFECTS OF LONG-TERM PRACTICE ON COORDINATION BETWEEN DIFFERENT JOINT MOTIONS IN STREET DANCERS

***Akito Miura*, PhD, Kazutoshi Kudo, PhD, Tatsuyuki Ohtsuki, PhD and Kimitaka Nakazawa, PhD***

Department of Life Sciences
Graduate School of Arts and Sciences, The University of Tokyo, Tokyo

## ABSTRACT

This study investigated practice-induced change in coordination among the movement of different joints in street dancers. It did so by quantifying similarities in shape and phase lag among hip, knee, and ankle joint angles. Skilled street dancers and non-dancers were asked to synchronize a standing position knee-bending movement to a metronome beat over a wide range of beat frequencies. The similarity in the shape between the time series of the joint angle pairs was characterized by the peak value of a cross correlation function (CCF).

The peak value of CCF in the hip-ankle angle pair was found to be significantly greater in dancers than in non-dancers, indicating coordination to be higher in dancers. The phase lag between hip and knee/ankle joint movements characterized by the time shift of the peak CCF value was found to be significantly greater in dancers than in non-dancers.

Because humans intrinsically have difficulties in developing a phase lag among joint movements, the greater extent of coordination and larger phase lag in dancers suggests that motor learning of coordination among joint movements in street dance is characterized by a stabilizing coordination pattern that involves an increase in phase lag among joint movements. This stabilizing pattern may play an important role in artistic expressions in street dance.

*Corresponding Author Email: akito.miura@gmail.com; Laboratory of Sports Sciences, address: Komaba 3-8-1, Meguro, Tokyo 153-8902, Japan.

**Keywords:** street dance, cross correlation function, coordination dynamics, motor learning, motor control

## INTRODUCTION

Street dance is a dance style originally developed by African-Americans (Shichirui, 1999). The population of street dancers has recently grown, and several international exhibitions and competitions are held every year. A salient feature of street dance is that essentially all music beats, which tend to be approximately 100 beats per minute (bpm), are coordinated and synchronized with movements. This has led to a belief that skilled street dancers who continue to practice a variety of dance elements over a period of years have superior abilities to coordinate whole-body movements with musical rhythms. Although many dance studies have examined ballet (see Hincapié, Morton, & Cassidy, 2008; Koutedakis, Stavropoulos-Kalinoglou, & Metsios, 2005; Wilson & Kwon, 2008 for review), there have been few investigations on the motor control of street dancers. Because coordinated whole-body movement to music is a common feature of many dance styles, we believe that an empirical study of street dancers can contribute to clarification of motor control not only of street dancers but also of other rhythmic dance styles. Further, it may also contribute to the development of effective training methods with regard to rhythmic dance.

Miura, Kudo, Ohtsuki, and Kanehisa (2011) described one of the salient properties of motor control in street dancers and reported that street dancers perform basic rhythmic knee-bending movements to a metronome beat with improved temporal accuracy as a result of long-term practice. Practice of a motor task leads to increased performance accuracy in temporal and spatial aspects (Sakurai & Ohtsuki, 2000; Schmidt & Lee, 2005). In addition to increasing performance accuracy, practice leads to a variety of changes in human movements, such as changes in the coordination pattern between different joint motions. In general, practice-induced changes in coordination patterns among joint movements have been investigated by quantifying similarity of shapes among the joint angle time series (McDonald, Van Emmerik, & Newell, 1989; Newell & Van Emmerik, 1989; Vereijken, Van Emmerik, Whiting, & Newell, 1992). For instance, Ko, Challis, and Newell (2003) reported that, in the early stage of practice of a dynamic balance task, similarities of shape existing among the time series of hip, knee, and ankle joint angles were low. With practice, these similarities increased, indicating that the coordination among joint movements can be improved by training (Ko, et al., 2003).

Skilled street dancers who continued to practice a variety of dance elements over a period of years showed temporally accurate performance during basic rhythmic knee-bending movements (Miura, et al., 2011; Miura, Kudo, Ohtsuki, Kanehisa, & Nakazawa, 2012). Considering the aforementioned research on increased coordination between joint motions by practice (Ko, et al., 2003), for street dancers to attain accurate performance, the degree of coordination among joint movements can be changed and greater similarities among time series of joint angles can be achieved through motor learning and training. The purpose of the current study was to investigate this hypothesis by calculating the cross correlation function (CCF) among joint angles for a basic rhythmic knee-bending movement. Skilled street dancers were examined and novices provided a control group. CCF can quantify both the degree of similarity and the temporal relationship (i.e., phase lag/relative phase) between rhythmic joint movements.

In order to quantify practice-induced coordination changes among joint movements in street dancers, not only the extent of similarity in shape but also the temporal relationship of joint motions should be investigated. The motor system has an intrinsic preference for a temporal relationship between rhythmic joint movements, which is derived from initial capacities before motor learning occurs (Zanone & Kelso, 1992). However, street dancers showed various temporal relationships between joints as a means of artistic expression. In human rhythmic movement, the movement of joint angles in the same phase relation (phase lag 0/relative phase 0°) is the most stable coordination pattern compared to other phase relations (Kelso, 1981, 1984, 1995; Kelso, Buchanan, & Wallace, 1991) because, with increasing movement frequency, different phase relations between joint movements (relative phases other than 0°) are unintentionally entrained to the most stable phase relation (phase lag 0/relative phase 0°). However, previous studies have demonstrated that after participants sufficiently practice a different phase relation (relative phases other than 0°) between rhythmic joint movements (in other words, they sufficiently practice making a given phase lag between joint motions) they do not obtain the automatic entrainment identified above (Zanone & Kelso, 1992, 1997). Thus they can preserve a phase lag between the joints during rhythmic movement. Because artistic expression encourages street dancers to practice the development of a phase lag between joint movements of arms, legs, and trunk, we hypothesized that the temporal relationship (phase lag/relative phase) between the joint movements will differ between street dancers and novice group. Further, we expect that the degree of coordination among joints, (i.e., the similarity in shapes among the time series of joint movements) will be higher in street dancers than in non-dancers.

# MATERIALS AND METHODS

## Participants

Eight skilled street dancers and 8 novice controls took part in this experiment. Hereafter, the dancer group is referred to simply as "dancers," and the control group, as "non-dancers." The dancers (men, 24.3 ± 4.4 years of age, mean ± standard deviation [SD]) had 7.3 ± 4.3 years of dancing experience and included 6 winners of celebrated national or international street dance competitions. Non-dancers (men, 23.9 ± 1.1 years of age) had no experience in any type of dance. All participants were free of musculoskeletal disorders. Informed consent was obtained from all participants and the study was approved by the Ethics Committee of the Graduate School of Arts and Sciences, the University of Tokyo.

## Experimental Task

The experimental task was a knee-bending movement, which constitutes a basic movement in street dance. The movement involves bouncing up and down repeatedly by bending at the knees in response to a metronome beat (Figure 1). The participants were instructed to bend their knees to the beat, and keep a 1:1 relationship between knee movement and beat. The beat rate ranged from 40 to 180 bpm in increments of 20 bpm. One beat rate was used for each trial. Participants were also instructed not to move the neck or parts of the body other than the hip, knee, and ankle joints.

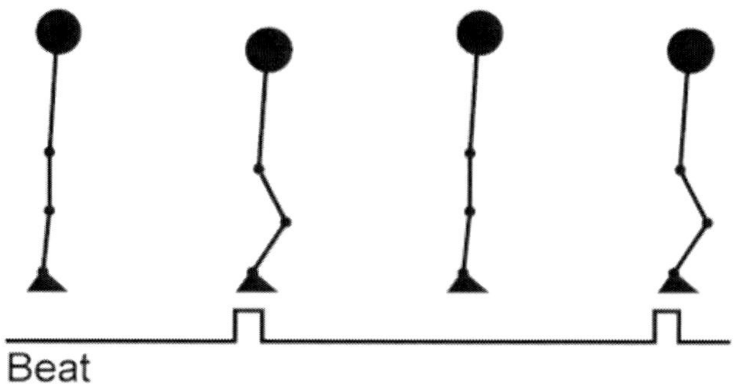


Figure 1. Experimental movement task flexes the knees on the beat.

## Design and Procedure

The participants performed eight trials (8 beat rates). Trial durations were approximately 30 s, 24 s of which were analyzed. The order of the trials was randomized. Immediately before each trial, all participants watched a short instructional video demonstrating the trial movement performed by a street dancer. During and after watching the video, the participants were allowed to practice the movement for a few minutes. The trials began after the short practice session. To exclude the influence of fatigue, the participants were allowed to take sufficient rest breaks between trials, for at least a few minutes or more, to the extent that they did not experience any fatigue or discomfort. Participants were instructed to try to coordinate movement to the beat while in motion, even if they could not perform the movement precisely to the beat.

## Apparatus and Data Collection

All trials were videotaped with a high-speed camera (HAS-220; DITECT, Tokyo, Japan) directed to the sagittal plane. Digital images were recorded at an acquisition rate of 200 Hz and all images were included in the analysis. Reflective markers were attached to the right side of the acromion, the greater trochanter, the rotation axis of the knee joint, the lateral malleolus, and the first metatarsophalangeal joint. The coordinates of all markers were digitized using motion analysis software (DIPP-Motion Pro; DITECT, Tokyo, Japan). To eliminate potential start-up effects, videotaping was begun as soon as each participant had performed approximately 10 cycles of the knee-bending movements.

## Data Analysis

Prior to analysis, three of the 128 trials (8 trials × 16 participants) were excluded because the participants of these 3 trials could not maintain a 1:1 relationship between the movement and the beat, or they were incapable of retaining the movement to the beat. Hip, knee, and ankle angular displacements were computed from the coordinates of the reflective markers. The angular displacement data were smoothed using a low-pass filter (cutoff frequency, 6 Hz).

Cross correlation functions (CCFs) among joint angle data were calculated. CCF quantifies the extent of similarity in shape and the temporal relationship (phase lag/phase relation) between 2 joint angle time series (for reference, see Nelson-Wong, Gregory, Winter, & Callaghan, 2008; Nelson-Wong, Howarth, Winter, & Callaghan, 2009). The peak value of the CCF quantifies the extent of similarity in shape between two joint angle time series in a manner that is independent of their temporal relationship. Where two joint angle time series have exactly the same shape, the maximal peak value of the CCF will be 1. If the peak CCF value is 0, it implies that the shapes of 2 joint angle time series have no similarity and that the 2 joint angles move independently of each other.

Time shifts of the peak value quantify the temporal relationship (phase lag/phase relation) between joint angle time series. For example, if the peak value of CCF is 1, and the time shift of the peak value is 5 ms, the 2 joint angle time series are identical in shape with a phase lag of 5 ms. We calculated peak values of CCF and the time shift of the peak values. Time shifts were normalized by the interstimulus interval (ISI).

## Statistical Analysis

First, missing data was replaced by the mean of all participants at the respective movement frequency. A 2-way analysis of variance (ANOVA) with one within-subject factor (beat rate; 40, 60, 80, 100, 120, 140, 160, and 180 bpm) and one between-subject factor (group; dancer vs. non-dancer) was then performed on the following: (1) the peak value of the CCF and (2) the time shift of the peak value normalized by the ISI. When Mauchly's test of sphericity showed heterogeneity of covariance in the ANOVA, the more conservative Greenhouse–Geisser test was performed. Statistical significance was set at $p < .05$.

# RESULTS

## Peak CCF Value

Figure 2 shows the mean peak CCF value in each joint pair as a function of beat rate. A 2-way ANOVA on the peak CCF value indicated a significant main effect of beat rate on the hip-knee pair, $F(1.80, 25.25) = 17.17$, $p < .01$, the ankle-knee pair, $F(1.459, 20.429) = 11.41$, $p < .01$, and the hip-ankle pair, $F(1.90, 26.59) = 19.63$, $p < .01$.

The significant main effect of group was found in the hip-ankle pair, $F(1, 14) = 6.25$, $p < .05$. No other main effects or interactions were found to be statistically significant. The significant main effect of group means that, for dancers, the peak CCF value for the hip-ankle pair was greater than that of non-dancers. This indicates that the joint angle time series between the hip and ankle for dancers has greater similarity in shape than for non-dancers.

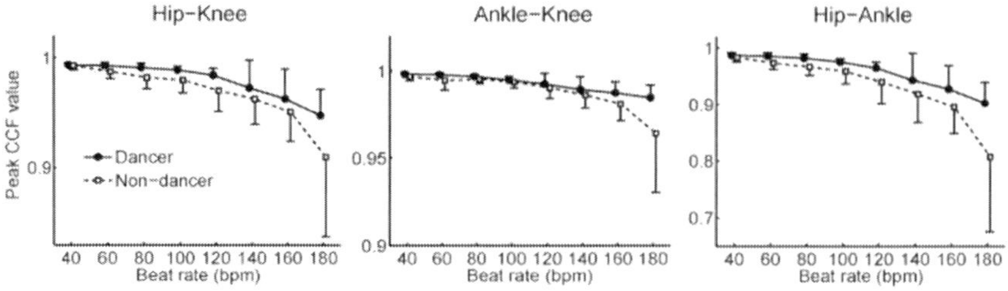


Figure 2. Peak CCF value for each joint angle pair. Vertical bars represent between-subject standard deviations.

## Time Shift of Peak CCF Value

Figure 3 shows the mean of the time shift of peak CCF values in each joint pair as a function of beat rate. In the 2-way ANOVA on the time shift of the peak CCF value, the significant main effect of the beat rate was found in the hip-knee pair, $F(2.43, 34.02) = 11.41$, $p < .01$ and the hip-ankle pair, $F(1.88, 26.29) = 8.14$, $p < .01$; the significant main effect of the group was found in the hip-knee pair, $F(1, 14) = 13.78$, $p < .01$ and the hip-ankle pair, $F(1, 14) = 5.57$, $p < .05$. No other main effects or interactions were found to be statistically significant. The significant group main effects mean that, for dancers, time shifts of the peak CCF values in hip-knee and hip-ankle pairs deviate from 0 to an extent greater than those of non-dancers. This indicated that the dancers had a greater phase lag between hip and knee/ankle joint angle time series. In other words, for dancers, the hip angle time series preceded the knee/ankle joint time series more than the phase lag observed for non-dancers.

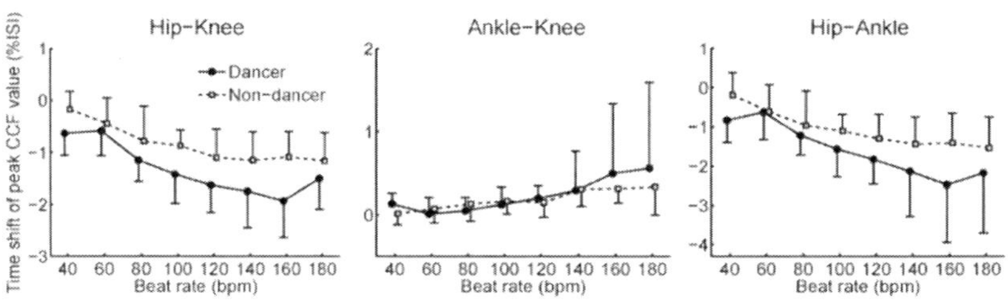


Figure 3. Time shift of the peak CCF value for each joint angle pair. Vertical bars represent between-subject standard deviations.

## DISCUSSION

This study investigated practice-induced changes of coordination between joint motions by calculating the CCF between joint angles. Dancers and non-dancers performed a basic street dance knee-bending movement to a metronome beat over a wide range of frequencies. Analysis of the CCF between the joint angles revealed significant differences between dancers and non-dancers with respect to both peak values and the time shifts of peak values.

This suggests that coordination patterns between joint movements had changed through long-term practice. We discuss this in terms of the intrinsic preference of the motor system for phase relation between joint motions.

The dancers were found to exhibit a significantly greater peak CCF value (index of similarity in shape of joint angle time series) in the hip-ankle pair than the non-dancers (Figure 2). The peak CCF value in the hip-knee and ankle-knee pairs in dancers also tended to be greater than that of the non-dancers ($p < .10$ and $p < .10$). These results indicate that the shape of the joint angle time series has greater similarity in dancers than in non-dancers. Thus, the results are consistent with a previous study wherein the results of a dynamic balance task indicated that similarity in the shape of joint angle time series becomes greater with practice (Ko, et al., 2003). The present results suggest that the similarity in the shape of joint angle time series can represent an index for the extent of coordination between joints during rhythmic multi-joint movement.

The time shift of the peak value that quantified the phase lag between time series of joint angles of dancers deviated significantly more from the phase lag 0 (relative phase 0°) in the hip-knee/ankle pairs than that of non-dancers (Figure 3). For dancers, the hip angle time series preceded the knee and ankle angle time series to a greater extent than for the non-dancers. In human movement coordination, the motor system has an intrinsic preference for a phase lag/phase relation between rhythmic joint motions. In other words, one can perform a certain joint coordination pattern stably without any practice, because of capacities existing before motor learning occurs (Zanone & Kelso, 1992). This phenomenon has been investigated within the framework of coordination dynamics. Stability of coordination pattern is characterized by relative phase variability and the direction of entrainment. When the period of rhythmic movement of a joint is expressed as 360 degrees, relative phase quantifies where the peak of another joint lies within the 360 degrees. Thus, the same phase relation is described as relative phase 0 or 360 degrees, and an exactly alternate phase relation as 180 degrees. In inter-limb coordination, relative phase variability is the lowest at same phase relation (relative phase 0°) and it is higher at alternate phase relation (relative phase 180°). It is highest at the intermediate phase relation between the same and alternate phase relations (Kelso, 1995). In addition, in wrist-elbow rhythmic coordination an alternate phase relation (wrist extension synchronized with elbow flexion and wrist flexion with elbow extension; relative phase 180°) was found to be suddenly replaced by same phase relation (wrist flexion synchronized with elbow flexion and wrist extension with elbow extension; relative phase 0°) with an increase in movement frequency (Kelso, et al., 1991). Thus, in terms of coordination dynamics, the same phase relation is described as an “intrinsically stable” phase relation because 1) one can perform it with small relative phase variability without any practice, and 2) this phase relation never be unintentionally entrained to other phase relations. On the other hand, an alternate phase relation is described as “intrinsically less stable”. Phase relation other than the intrinsically stable or less stable ones is described as “intrinsically unstable”. These are because 1) during performance of these phase relations, relative phase variability becomes greater and 2) these phase relations are automatically and unintentionally entrained to the same phase relation unless sufficient practice is done. This has been confirmed in various types of coordination regardless of what joints are involved, such as in single arm multi-joint coordination (Buchanan, Kelso, & deGuzman, 1997), bimanual finger coordination (Haken, Kelso, & Bunz, 1985; Kelso, 1981, 1984; Schoner, Haken, & Kelso, 1986), inter-personal finger coordination (Oullier, deGuzman, Jantzen, Lagarde, & Kelso, 2008) and inter-personal leg coordination (Schmidt, Carello, & Turvey, 1990). Considering these research, phase relation of non-dancers who had never practiced the task may be intrinsically stable phase relations, which can be performed without any practice.

However, research has shown that, with practice, one can stably continue phase relations other than the same phase relation. In other words, phase relations do not show automatic entrainment to the same phase relation with sufficient practice of the phase relation (Zanone & Kelso, 1992, 1997). This has also been confirmed in various types of coordination regardless of what joints are involved, such as in single arm multi-limb coordination (Buchanan, 2004; Buchanan, Zihlman, Ryu, & Wright, 2007), bimanual finger coordination (Zanone & Kelso, 1992, 1997), bimanual arm coordination (Fontaine, Lee, & Swinnen, 1997; Fujii, Kudo, Ohtsuki, & Oda, 2010; Kostrubiec & Zanone, 2002; Lee, Swinnen, & Verschueren, 1995; Smethurst & Carson, 2001; Swinnen, Lee, Verschueren, Serrien, & Bogaerds, 1997; Tallet, Kostrubiec, & Zanone, 2008; Wenderoth, Bock, & Krohn, 2002) and both legs coordination (Kelso & Zanone, 2002).

In the current study, a comparison of subject groups showed dancers to exhibit a significantly greater deviation from the same phase relation (i.e., phase lag 0/relative phase 0°) between hip and knee/ankle angles (Figure 3) with a greater coordination between joint motions (the higher peak CCF value) (Figure 2) than for non-dancers. This suggests that dancers performed intrinsically unstable phase relation (one that is different from that of non-dancers) with greater degrees of coordination through modification of an intrinsically unstable phase relation into a stable one as a result of long-term practice.

Bernstein (1967) expressed practice-induced changes of coordination pattern between joint motions as the reorganization of biomechanical degrees of freedom. This means that the motor system has an intrinsically preferred relationship between joint motions and, with practice, relationships between joint motions become reorganized or changed. Thus, the reorganization of biomechanical degrees of freedom in street dancers can be characterized as stabilization of a new coordination pattern, one which involves an increase in the phase lag between joint motions. Because the present study is a comparison study between dancers and non-dancers, it is likely that those who have the innate ability to produce a phase lag between joint motions could be successful street dancers. However, considering previous research documenting the observation that phase relations other than intrinsically stable ones become stable with practice, and that skilled and complex movement of musicians is acquired by motor learning (Münte, Altenmüller, & Jäncke, 2002), the ability of street dancers to develop a phase lag between joint motions could be developed by long-term practice. This may be confirmed through future longitudinal studies.

The results of the present study together with the results of a previous study on action-perception coordination in street dancers (Miura, et al., 2011; Miura, Kudo, & Nakazawa, 2013), in which the intrinsically less stable action-perception phase relation was stabilized in street dancers, provide support for the proposal that, in terms of coordination dynamics, motor learning in street dance can be considered to be a process of stabilizing or overcoming an intrinsically less stable or unstable coordination patterns derived from intrinsic constraints imposed on the spatiotemporal organization of coordinated movement (Fujii, et al., 2010; Swinnen, 2002). Low variation in coordination patterns and stereotyped movements in beginners of street dance may stem from an intrinsic difficulty in developing a phase lag between joint motions. In contrast, in skilled street dancers, the higher variation of coordination patterns and diverse and creative movements, which has been suggested to appear with practice (Kudo & Ohtsuki, 2008; Kudo, Tsutsui, Ishikura, Ito, & Yamamoto, 2000), would be acquired by modifying such intrinsically less stable or unstable phase relations into stable relations. Because a phase lag between joint angles such as those in the arms, legs, and trunk is important for impressive and artistic expressions in street dance, a greater phase lag between joint motions in street dancers may be cultivated to pursue such expressions.

## CONCLUSION

We underscore the fact that although there has been little research on street dancers (Miura, et al., 2011; Miura, et al., 2013; Miura, et al., 2012), studies of accomplished street dancers can provide knowledge on motor skills performing street dance, which should contribute to the development of effective training and educational methods in whole-body rhythmic dance. The present study confirms that, although producing phase lags between joint motions is intrinsically difficult without practice, dancers developed more phase lag between joint motions than did non-dancers. Practice making phase lags between various joint motions would be beneficial for skill improvement in street dance.

## ACKNOWLEDGMENTS

This study was presented at the 20th Annual Meeting of the International Association for Dance Medicine and Science held in Birmingham, UK in October 2010. This study was partly supported by a Grant-in-Aid for Scientific Research (No.21300215 and No. 20240060) from Japan Society for the Promotion of Science (JSPS) awarded to K. Kudo, a Grant-in-Aid for Scientific Research (B) (No.19300216) from JSPS awarded to T. Ohtsuki, a Grant-in-Aid for JSPS Fellows (No.23· 9480) awarded to A. Miura, and Sports Research Grant from Yamaha Motor Foundation for Sports awarded to A. Miura.

## REFERENCES

Bernstein, N. A. (1967). *The co-ordination and regulation of movements*: Pergamon.

Buchanan, J. J. (2004). Learning a single limb multijoint coordination pattern: the impact of a mechanical constraint on the coordination dynamics of learning and transfer. *Experimental Brain Research*, 156(1), 39-54.

Buchanan, J. J., Kelso, J. A. S., & deGuzman, G. C. (1997). Self-organization of trajectory formation .1. Experimental evidence. *Biological Cybernetics*, 76(4), 257-273.

Buchanan, J. J., Zihlman, K., Ryu, Y. U., & Wright, D. L. (2007). Learning and transfer of a relative phase pattern and a joint amplitude ratio in a rhythmic multijoint arm movement. *Journal of Motor Behavior*, 39(1), 49-67.

Fontaine, R. J., Lee, T. D., & Swinnen, S. P. (1997). Learning a new bimanual coordination pattern: Reciprocal influences of intrinsic and to-be-learned patterns. *Canadian Journal of Experimental Psychology*, 51(1), 1-9.

Fujii, S., Kudo, K., Ohtsuki, T., & Oda, S. (2010). Intrinsic constraint of asymmetry acting as a control parameter on rapid, rhythmic bimanual coordination: A study of professional drummers and non-drummers. *Journal of Neurophysiology*, 104(4), 2178-2186.

Haken, H., Kelso, J. A. S., & Bunz, H. (1985). A theoretical-model of phase-transitions in human hand movements. *Biological Cybernetics*, 51(5), 347-356.

Hincapié, C. A., Morton, E. J., & Cassidy, J. D. (2008). Musculoskeletal injuries and pain in dancers: a systematic review. *Archives of Physical Medicine and Rehabilitation*, 89(9), 1819-1829.

Kelso, J. A. S. (1981). On the oscillatory basis of movement. *Bulletin of the Psychonomic Society,* 18(2), 63-63.

Kelso, J. A. S. (1984). Phase-transitions and critical-behavior in human bimanual coordination. *American Journal of Physiology*, 246(6), 1000-1004.

Kelso, J. A. S. (1995). *Dynamic patterns: The self-organization of brain and behavior*: The MIT Press.

Kelso, J. A. S., Buchanan, J. J., & Wallace, S. A. (1991). Order parameters for the neural organization of single, multijoint limb movement patterns. *Experimental Brain Research*, 85(2), 432-444.

Kelso, J. A. S., & Zanone, P. G. (2002). Coordination dynamics of learning and transfer across different effector systems. *Journal of Experimental Psychology-Human Perception and Performance*, 28(4), 776-797.

Ko, Y., Challis, J., & Newell, K. (2003). Learning to coordinate redundant degrees of freedom in a dynamic balance task. *Human Movement Science,* 22(1), 47-66.

Kostrubiec, V., & Zanone, P. G. (2002). Memory dynamics: Distance between the new task and existing behavioural patterns affects learning and interference in bimanual coordination in humans. *Neuroscience Letters*, 331(3), 193-197.

Koutedakis, Y., Stavropoulos-Kalinoglou, A., & Metsios, G. (2005). The significance of muscular strength in dance. *Journal of Dance Medicine & Science*, 9(1), 29-34.

Kudo, K., & Ohtsuki, T. (2008). Adaptive variability in skilled human movements. *Information and Media Technologies*, 3(2), 409-420.

Kudo, K., Tsutsui, S., Ishikura, T., Ito, T., & Yamamoto, Y. (2000). Compensatory coordination of release parameters in a throwing task. *Journal of Motor Behavior*, 32(4), 337-345.

Lee, T. D., Swinnen, S. P., & Verschueren, S. (1995). Relative phase alterations during bimanual skill acquisition. *Journal of Motor Behavior*, 27(3), 263-274.

McDonald, P., Van Emmerik, R., & Newell, K. (1989). The effects of practice on limb kinematics in a throwing task. *Journal of Motor Behavior,* 21(3), 245-264.

Miura, A., Kudo, K., Ohtsuki, T., & Kanehisa, H. (2011). Coordination modes in sensorimotor synchronization of whole-body movement: A study of street dancers and non-dancers. *Human Movement Science,* 30(6), 1260-1271.

Miura, A., Kudo, K., & Nakazawa, K. (2013). Action-perception coordination dynamics of whole-body rhythmic movement in stance: A comparison study of street dancers and non-dancers. *Neurosience Letters*.544, 157-162.

Miura, A., Kudo, K., Ohtsuki, T., Kanehisa, H., & Nakazawa, K. (2012). Relationship between muscle co-contraction and proficiency in whole-body sensorimotor synchronization: A comparison study of street dancers and non-dancers. *Motor Control*. 17(1), 18-33.

Münte, T. F., Altenmüller, E., & Jäncke, L. (2002). The musician's brain as a model of neuroplasticity. *Nature Reviews Neuroscience*, 3(6), 473-478.

Nelson-Wong, E., Gregory, D. E., Winter, D. A., & Callaghan, J. P. (2008). Gluteus medius muscle activation patterns as a predictor of low back pain during standing. *Clinical Biomechanics,* 23(5), 545-553.

Nelson-Wong, E., Howarth, S., Winter, D. A., & Callaghan, J. P. (2009). Application of autocorrelation and cross-correlation analyses in human movement and rehabilitation research. *The Journal of orthopaedic and sports physical therapy*, 39(4), 287-295.

Newell, K., & Van Emmerik, R. (1989). The acquisition of coordination: preliminary analysis of learning to write. *Human Movement Science*, 8(1), 17-32.

Oullier, O., deGuzman, G. C., Jantzen, K. J., Lagarde, J., & Kelso, J. A. S. (2008). Social coordination dynamics: Measuring human bonding. *Social neuroscience*, 3(2), 178-192.

Sakurai, S., & Ohtsuki, T. (2000). Muscle activity and accuracy of performance of the smash stroke in badminton with reference to skill and practice. *Journal of Sports Sciences*, 18(11), 901-914.

Schmidt, R. A, & Lee, T. D. (2005). *Motor control and learning: A behavioral emphasis.* Champaign, IL: Human Kinetics Publishers.

Schmidt, R. C., Carello, C., & Turvey, M. T. (1990). Phase transitions and critical fluctuations in the visual coordination of rhythmic movements between people. Journal of Experimental Psychology: *Human Perception and Performance*, 16(2), 227-247.

Schoner, G., Haken, H., & Kelso, J. A. S. (1986). A stochastic-theory of phase-transitions in human hand movement. *Biological Cybernetics*, 53(4), 247-257.

Shichirui, S. (1999). *Kokujinrizumukannohimitsu* [The secret of the sense of rhythm in African-American]. Tokyo: Ikuhousha.

Smethurst, C. J., & Carson, R. G. (2001). The acquisition of movement skills: Practice enhances the dynamic stability of bimanual coordination. *Human Movement Science*, 20(4-5), 499-529.

Swinnen, S. P. (2002). Intermanual coordination: from behavioural principles to neural-network interactions. *Nature Reviews Neuroscience*, 3(5), 348-359.

Swinnen, S. P., Lee, T. D., Verschueren, S., Serrien, D. J., & Bogaerds, H. (1997). Interlimb coordination: Learning and transfer under different feedback conditions. *Human Movement Science,* 16(6), 749-785.

Tallet, J., Kostrubiec, V., & Zanone, P. G. (2008). The role of stability in the dynamics of learning, memorizing, and forgetting new coordination patterns. *Journal of Motor Behavior,* 40(2), 103-116.

Vereijken, B., Van Emmerik, R., Whiting, H., & Newell, K. (1992). Free (z) ing degrees of freedom in skill acquisition. *Journal of Motor Behavior*, 24(1), 133-142.

Wenderoth, N., Bock, O., & Krohn, R. (2002). Learning a new bimanual coordination pattern is influenced by existing attractors. *Motor Control*, 6(2), 166-182.

Wilson, M., & Kwon, Y. H. (2008). The role of biomechanics in understanding dance movement: A review. *Journal of Dance Medicine & Science*, 12(3), 109-116.

Zanone, P. G., & Kelso, J. A. S. (1992). Evolution of behavioral attractors with learning - Nonequilibrium phase-transitions. *Journal of Experimental Psychology-Human Perception and Performance,* 18(2), 403-421.

Zanone, P. G., & Kelso, J. A. S. (1997). Coordination dynamics of learning and transfer: Collective and component levels. *Journal of Experimental Psychology-Human Perception and Performance,* 23(5), 1454-1480.

In: Advances in Arts Biomechanics
Editors: Gongbing Shan and Peter Visentin

ISBN: 978-1-53615-641-6


*Chapter 9*

# THE INFLUENCE OF WARM-UP ON BALLET DANCERS' RISK OF INJURY

***Amber Mueller****
Pharmacy Program
University of Alberta, Edmonton, Canada

## ABSTRACT

This study explored the effects of warm-up on ballet dancers' risk of injury. Specifically, it looked at the influence of warm-up on range of motion (ROM) and muscle activity using electromyography (EMG) of the dancers' lower extremities. A convenient sample of 2 female ballet dancers, ages 15 and 21 years, participated in this study. The dancers took part in a short ballet class and performed several ballet techniques which were measured at 0%, 50%, 75%, and 100% of the warm-up cycle. 3D motion capture and biomechanical modeling were used during each trial to calculate and synchronize ROMs and EMG levels of selected muscles and joints of the legs. Results demonstrated that, on average, warm-up causes an increase in ROM and a decrease in EMG activity. From this, it is reasonable to extrapolate that preliminary exercise is important to prepare the body for more vigorous exercise and this may consequently be a useful tool for dance instructors to help reduce injury rates among dancers.

**Keywords:** 3D motion capture, range of motion, electromyography, dance, injury

## INTRODUCTION

Ballet is a graceful and elegant art form that has existed since the 14th century. It has all the "elements of the arts in its make up – drama, poetry, literature, painting, sculpture, design, music, and of course, dance" (Kennedy, Hodgkins, Colombier, Guyette, & Hamilton, 2007, p. 142). Experiencing a ballet performance can be inspiring for both the performers and the

* Address of corresponding author: Amber Mueller, Pharmacy Program, University of Alberta, 116 St. and 85 Ave., Edmonton, AB, Canada T6G 2R3. E-mail: amber.mueller@uleth.ca

audience as the "choreographer sculpts a composition of expressive motion" using ballet dancers' bodies and movements (Kennedy et al., 2007, p. 142). The dancers have a way of making each movement meaningful and they appear to do so with ease. But, mastering the skills needed in ballet requires repetitive practice and long-term dedication. The vigorous physical training endured by ballet dancers during learning and performance does not come without costs.

Ballet, being a performance art, places a great emphasis on aesthetics. This may lead to dancers having poor body image, which may lead to a variety of problems from eating disorders to musculo-skeletal pathologies. Professional stress is a "given" because obtaining a job as a professional ballet dancer is highly competitive, the income is very low, and dancers often "believe that dance perfection is more important than their own physical" and emotional well-being (Shah, 2008, p. 130). Also, because ballet "is not only a performing art, but a highly rigorous athletic sport," dancers often succumb to injury during practice or performance (Shah, 2008, p. 128). It has been "suggested that ballet dancers are at increased risk of injury because of the tremendous physical and emotional demands placed upon them" (Adam, Brassington, Steiner, & Matheson, 2004, p. 43). Many choreographies force dancers to test their physical limits. The end result may look very beautiful on stage, but it can take a toll on the dancer's body.

Studies have shown that between 85%-95% of professional ballet dancers suffer from at least one self-reported vocational injury (Twitchett et al., 2010; Shah, 2008). Self-reported injury can be defined as "a physical problem deriving from stress or other causes to do with performance, rehearsal, training, touring or the circumstances of dance life, which affects your ability to participate fully in normal training, performance or physical activity" (Twitchett et al., 2010, p. 26). Dancers participating in different genres of dance often suffer from different kinds of injuries "depending on where their specific dance [style] puts strain on" the body (Shan, 2005, p. 64). Previous research indicates that "64%-80% of ballet injuries are in the lower extremities" (Shan, 2005, p. 64). This includes the foot, ankle, knee, and lower back. It is thought that the lower limbs suffer the most for female dancers who perform classical ballet (ballerinas) because of "the enormous strain the muscles and joints are put under to achieve certain ballet positions", such as turn out (or extreme external rotation) of the hips (Shan, 2005, p. 65). Other mechanisms of dance injuries can include lack of "adequate strength, flexibility, or technique," (Shah, 2008, p. 129) "overwork, unsuitable floors, difficult choreography, and insufficient warm-up" (Koutedakis & Jamurtas, 2004, p. 655). So, because of the shockingly high percentage of ballet dancers who are injured every year, it is timely for researchers to concentrate on determining ways to reduce dancers' risk of injury. The present study was designed to explore a means to do just that - reduce injury risk in ballet.

Previous research has suggested various ways to reduce injury in ballet dancers. It has been found that dancing on suspended floors helps to reduce injury in dancers because it decreases the shock and pressure in the legs (McGuinness & Doody, 2006). This helps to decrease knee and ankle injuries. Furthermore, the style of shoes that dancers wear can have an effect on injury rates. Dancers who wear split sole runners during rehearsal have a reduced injury risk (McGuinness & Doody, 2006). It has also been suggested that fatigue is a major "injury risk factor in dance" (Twitchett et al., 2010, p. 29). It was found that dancers who have low aerobic fitness levels often experience fatigue more rapidly (Twitchett et al., 2010). Once the dancer is "fatigued, the ability to perform movements requiring complex skill is compromised" and "faulty alignment, inefficient biomechanics, and stress" may result in the dancer becoming injured (Twitchett et al., 2010, p. 29). Also, Twitchett et al. (2010) discovered that dancers with low body weight are at an increased risk of injury. In the real world, some of the solutions suggested or implied in these researchers' studies might not be available to all ballet dancers. Not all studios or theaters are equipped with suspended flooring and it is very unlikely that ballerinas can wear split sole runners to ballet classes because, although they may mitigate risk for some exercises, they may impede movement for others. Therefore, it remains relevant to explore a variety of options to reduce injury risk among ballet dancers. The research question of this paper is: could sufficient levels of warm-up be the answer?

McGuinness and Doody (2006) looked at "the rate and types of injury in competitive" Irish dancers (p. 35). Ballet and Irish dance have similar characteristics. They both "involve turnout or external rotation of the hip", and the ankle and foot are the most common sites of injury (McGuinness & Doody, 2006, p. 37). This study found that a "preventative warm-up program used with professional Irish dancers [decreased] … injury rates from 60% to 20% over a 15 month period" (McGuinness & Doody, 2006, p. 38). Furthermore, in modern dance it was found that "a preventative program … reduces injury rates from 81% to 76% over a three year period" (McGuinness & Doody, 2006, p. 38).

Further research done by Malliou, Rokka, Beneka, Mavridis, and Godolias (2007) found that dance aerobic instructors' rate of injury is statistically related to "the duration of the warm up and cool down" (p. 33). It was found that when "the warm up and cool down … is about 15 minutes, the number of injured instructors appears significantly smaller" (Malliou et al., 2007, p. 33). Furthermore, a "private warm up and cool down before and after class, decreases the rate of injuries even more" (Malliou et al., 2007, p. 33). Malliou et al. (2007) believe that "the lack of flexibility creates injury conditions, while the appropriate warm up and cool down prepares the muscle groups for the different types of aerobic dance" (p.33).

Warm-up or preliminary exercise is an important aspect of most sports because it helps to prepare the athlete "physiologically or psychologically and reduces the likelihood of joint and muscle injury" (McArdle, Katch, & Katch, 2007, p. 588). A "warm-up provides a comfortable lead in to more vigorous exercise", "should activate specific muscles in a way that mimics the anticipated activity, … brings about full range of joint motion", and helps to increase the efficiency of contracting muscles (McArdle et al., 2007, p. 589). Warm-up is an important aspect of ballet class. It "makes the muscles more pliable, coordinated, responsive, and resistant to injury" (Gaynor Minden, 2005, p. 108). It often includes stretching, as well as doing basic ballet movements.

A typical ballet class begins "with dancers lightly holding onto the ballet barre for support" (Shah, 2008, p. 130) while they do exercises consisting of pliépliésspliès ("a bending movement of the knees" (Mara, 1987, p. 90)), battement tendus, rises, grand battements (big kick), and stretching. Barre exercises are important to warm-up the dancers' muscles for more difficult center movements that are to follow. Once the ballet instructor deems the dancers to be sufficiently "warmed-up", the class proceeds with center work. This includes port de bras ("carriage of the arms" (Mara, 1987, p. 91)), sautés (a small jump), grand jetés (big jump that resembles the dancer doing the splits in the air), and pirouettes (a "complete turn of the body on one foot" (Mara, 1987, p. 89)). Finally, the class ends with pointe work, where the dancer's "body weight is carried on the tips of the toes, primarily on the great toe" (Shah, 2008, p. 129). The dancer typically does bourrées/couru (quick walking-like movements en pointe), relevés (rising up onto the tips of the toes), grand battements, and pirouettes en pointe. Most ballet classes last 1-2 hours and professional dancers may train for 5-10 hours a day, 6 days a week (Haight, 1998).

As a rehearsal progresses, it gradually increases in intensity. This is because the dancer's joints need to be sufficiently warmed-up before they advance to more strenuous exercises in the center. Since warm-up is believed to help decrease the chance of injury in Irish dancers and aerobic dance instructors, could warm-up also help reduce injury risk in ballet dancers? Because there is a lack of research in this area, this present study used biomechanical modeling and analysis to detect movements that may cause injury. It specifically looked at the influence of warm-up on various ballet techniques, including grand battement, sauté, and grand jeté. This study looked at changes in range of motion (ROM) and electromyography (EMG) as the dancer became more warmed-up. ROM or joint flexibility is defined as the "angle through which a joint moves from anatomical position to the extreme limit of segment motion in a particular direction" and "is measured directionally in units of degrees" (Hall, 2007, p. 127). EMG "is used to study neuromuscular function, including identification of which muscles develop tension throughout a movement" and is measured in units of millivolts (Hall, 2007, p. 75). The technology can also be applied in studies of arts performance (Visentin, P. & Shan, G., 2011). It is hypothesized that as the dancer becomes more warmed-up, ranges of motion (ROMs) will increase. At peak ROM, the dancer should have a reduced risk of injury. Further, it is speculated that EMG will, at this point, decrease as muscle activity becomes more efficient and the muscles strain less.

## METHODS

### Participants

A convenient sample of 2 female dancers were selected to participate in this study. Subject A was 15 years old at the time of the study. She has been training in ballet for 9-10 years. Subject B was 21 years old at the time of the study and has been training in ballet for 13-14 years. Although the Subjects do not train at the same dance studio, they have received ballet training based on similar approaches. Before participating, each subject was informed of the purpose and goals of this study and they both voluntarily signed an informed consent. The subjects wore their own ballet attire and shoes for all trials of the study.

### Materials and Equipment

At the beginning of the session, each subject was dressed up in a skin-tight body suit and 41 reflective markers were placed on different parts of the subjects' body, including the anterior superior iliac crest, posterior superior iliac crest, lateral condyle of the tibia, lateral malleolous of the fibula, calcaneal tuberosity, the head of the hallicus, the upper and lower legs, acromion process, lateral epicondyle of the humerus, styloid process of the ulna and radius, third metacarpophalangeal joint, upper and lower arm, sternal notch, xiphoid process, C7, T10, right back, and head. A 12-camera VICON motion capture system (VICON Motion Systems, Oxford Metrics Ltd., Oxford, England) was used to capture and synchronize ballet movements with force plate data by tracking all 41 reflective markers. The VICON software triangulated positions of each marker and rendered a 3D model consisting of 15 segments (Shan & Westernhoff, 2005). In such biomechanical modeling, inertial characteristics of the body are estimated using anthropometric "norms" found through statistical studies (Shan & Bohn, 2003; Winter, 1990). As well, a Sony Digital camera was used to capture video/visual reference data.

Surface electrodes were placed onto the subject to capture EMG signals. Muscle activity was recorded for each of the subjects' quadriceps, hamstrings, and gastrocnemius muscles (both legs). Later, this data was analyzed to find changes in efficiency of each muscle during different stages of the warm-up process. Additionally, a force plate was used to measure the ground reaction forces (the force plate results will be discussed in a further study).

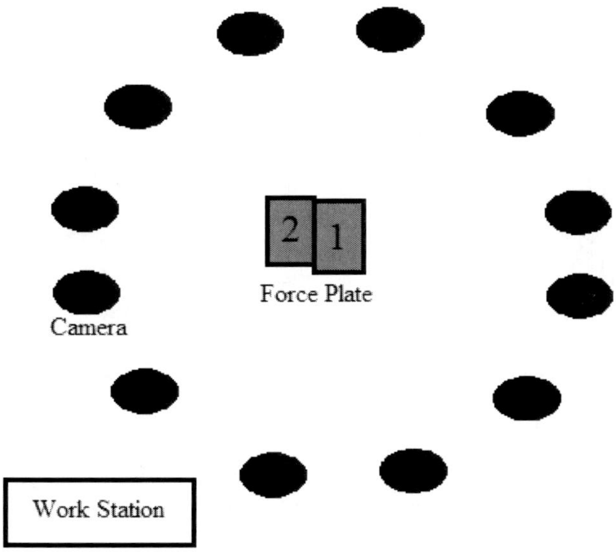


Figure 1. Schematic of the experimental setup including: 12 cameras, force platforms, and the computer work-station.

## Procedures

For the first trial the dancer was 0% warmed-up and performed three grand battements to the front, three to the side, three to the back, and three to the side with the left leg and was then repeated with the right leg (Figure 2A, B, C). The dancer held onto a barre while performing these techniques.

An experienced ballet instructor then ran a ballet class to warm-up the dancer to an estimated 50% of a full warm-up.

This short class consisted of various exercises that were designed to help the dancer warm-up specifically for grand battement and grand jeté. Some of the exercises included pliépliés, battement tendus, glissés, grand battement, and port de bras.

Once the dancer was 50% warmed-up, another data gathering session was performed. For this session, she performed three grand battements to the front, side, back, and side with both legs (Figure 2A, B, C); three sautés in first position (Figure 2E); and three grand jetés with the preparation of the dancer's choice (Figure 2F). Following this motion capture, the teacher instructed some more ballet exercises to warm-up the dancer to 75%. At this point, the dancer performed another capture session which consisted of three grand battements to the front, side, back, and side (Figure 2A, B, C); three sautés in first position (Figure 2E); and three grand jetés (Figure 2F).

The dancer was then instructed by the professional ballet trainer to perform jumping exercises and *relevés* in the center of the studio (without the barre) until she was 100% warmed-up. A data capture was repeated and the dancer performed the grand battements (*Figure 2A, B, C*), sautés (Figure 2E), and grand jetés (Figure 2F).

The dancer then put on her pointe shoes and performed another motion capture which consisted of three grand battements to the front, side, back, and side with both legs (*Figure 2A, B, C*); three *relevés* in fifth position (right foot forward); bourees going side to side on the force plate; and pirouette en dehors.

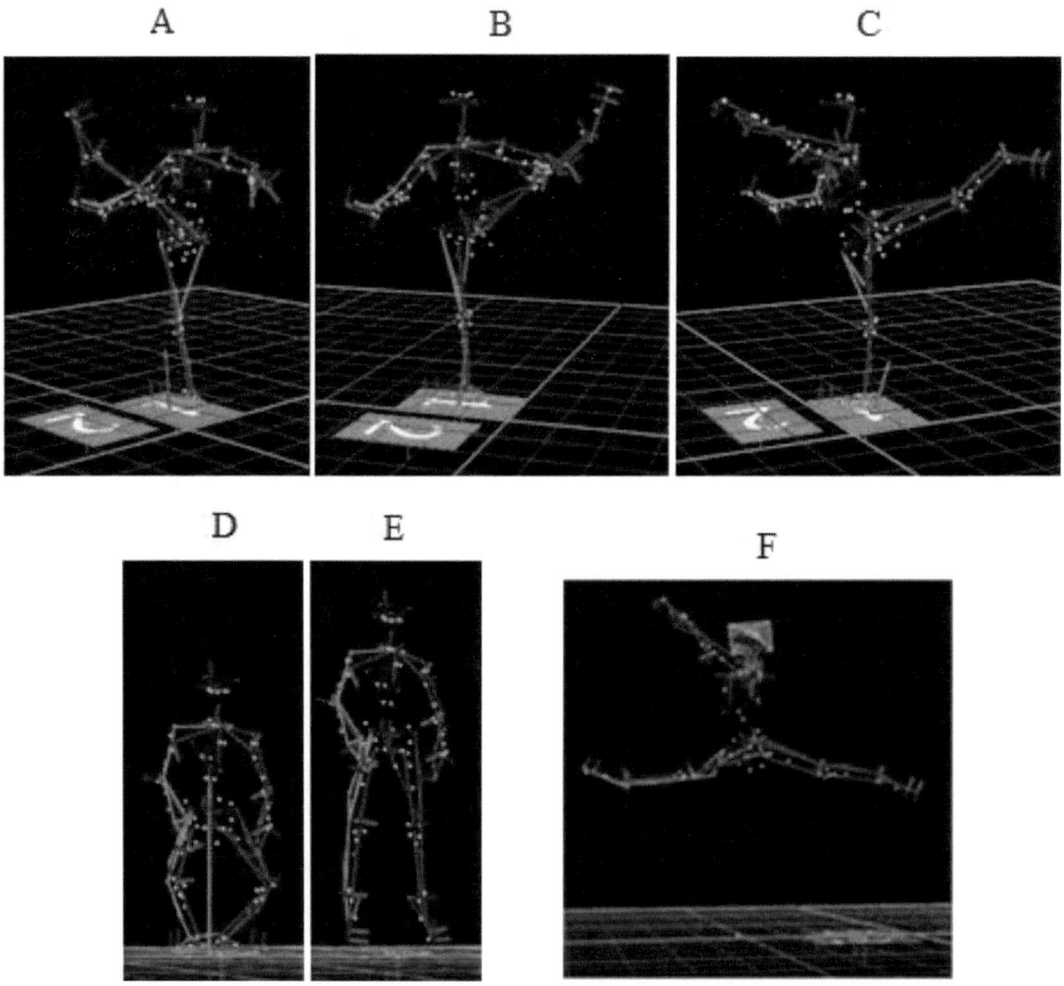


Figure 2. Model representation of grand battement to the front (A), to the side (B), and to the back (C); plié (D) and sauté (E) in first position with arms in bras bas; and grand jeté (F) with right foot in front and arms in second arabesque.

## Data Collection

Once the data was gathered, it was processed using VICON Life Sciences software. The processed data was then exported to Microsoft Excel where it was used to calculate joint angles or ranges of motion (ROMs) of the hip joints, knee joints, and ankle joints. ROM data includes three parameters: Flexion/extension, abduction/adduction, and rotation. Figure 3 describes each of these movements. MatLabsoftware was used to envelope raw EMG data in order to calculate the maximum EMG values of the hamstrings, quadriceps, and gastrocnemius muscles during each trial. Together, these data can be used to evaluate the efficiency of the muscles' activity.

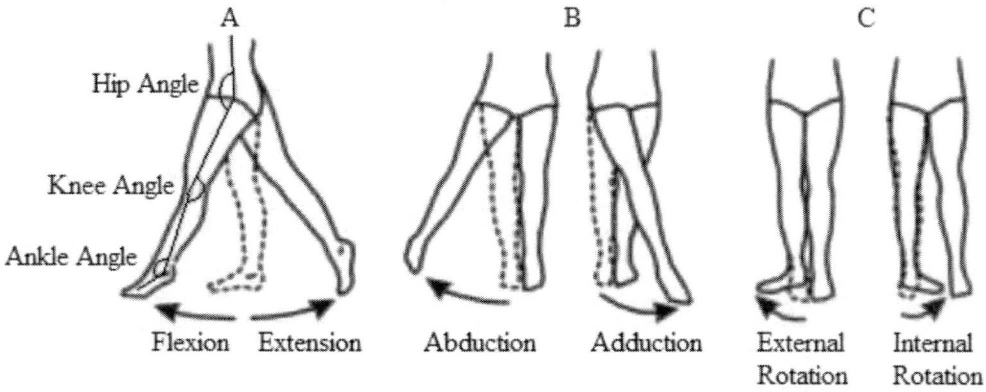


Figure 3. Schematic of flexion/extension, abduction/adduction, and external/internal rotation.

# RESULTS

## Observations

A qualitative professional evaluation of the subjects was performed and the several phenomena were identified. Subject A did two quick plié, instead of a single plié, in between consecutive sauté jumps during the pre warm-up. But, this seemed to decrease as she became more warmed-up. During the grand jeté, Subject A's preparation for the jump seemed to be more frantic when she was not warmed-up. As warm-up increased, the preparation was more controlled. Subject B lead with her toes during grand battement. Right before the kick, her leg internally rotated resulting in a less turned out leg. Subject B's grand jeté height was also visibly higher as she became more warmed-up. Furthermore, it was noticeable that as both subjects became more warmed-up their flexibility increased and the precision of their movements improved.

## Kinematics - Range of Motion

The following tables show the effect that warm-up had on the hip, knee, and ankle joints ROM, including flexion/extension, abduction/adduction, and rotation. The bold numbers indicate peak values for ROM.

**Table 1. Average ROM in degrees for kicking leg during left grand battement to the front during various degrees of warm-up (pre warm-up, 50%, 75%, 100%, and pointe)**

| | | Pre Warm-Up (°) | 50% (°) | 75% (°) | 100% (°) | Pointe (°) |
|---|---|---|---|---|---|---|
| Hip | Flexion/Extension | 107.75 | 110.14 | 106.72 | 108.87 | 107.05 |
| | Abduction/Adduction | 29.90 | 32.09 | 32.93 | 35.96 | 32.80 |
| | Rotation | 25.68 | 26.67 | 25.88 | 25.34 | 44.80 |
| Knee | Flexion/Extension | 54.44 | 61.89 | 65.47 | 67.22 | 65.15 |
| | Abduction/Adduction | 14.37 | 23.10 | 15.49 | 20.47 | 50.01 |
| | Rotation | 28.54 | 35.05 | 32.30 | 37.93 | 32.04 |
| Ankle | Flexion/Extension | 77.62 | 82.79 | 80.20 | 81.54 | 98.02 |
| | Abduction/Adduction | 45.86 | 41.81 | 39.51 | 38.46 | 33.87 |
| | Rotation | 45.79 | 61.73 | 50.70 | 59.57 | 54.10 |

**Table 2. Average ROM in degrees for kicking leg during left grand battement to the side during various degrees of warm-up (pre warm-up, 50%, 75%, 100%, and pointe)**

| | | Pre Warm-Up (°) | 50% (°) | 75% (°) | 100% (°) | Pointe (°) |
|---|---|---|---|---|---|---|
| Hip | Flexion/Extension | 116.06 | 128.88 | 124.78 | 125.37 | 116.07 |
| | Abduction/Adduction | 79.36 | 77.96 | 78.90 | 79.49 | 78.50 |
| | Rotation | 63.17 | 70.97 | 68.53 | 69.35 | 67.31 |
| Knee | Flexion/Extension | 69.19 | 61.33 | 63.45 | 82.32 | 76.45 |
| | Abduction/Adduction | 18.30 | 20.00 | 20.94 | 22.29 | 25.55 |
| | Rotation | 39.71 | 35.50 | 36.18 | 40.28 | 42.83 |
| Ankle | Flexion/Extension | 82.88 | 89.88 | 89.14 | 82.74 | 105.82 |
| | Abduction/Adduction | 46.37 | 52.49 | 44.53 | 42.68 | 48.85 |
| | Rotation | 62.92 | 68.82 | 60.74 | 65.56 | 56.99 |

**Table 3. Average ROM in degrees for kicking leg during left grand battement to the back during various degrees of warm-up (pre warm-up, 50%, 75%, 100%, and pointe)**

| | | Pre Warm-Up (°) | 50% (°) | 75% (°) | 100% (°) | Pointe (°) |
|---|---|---|---|---|---|---|
| Hip | Flexion/Extension | 78.44 | 87.81 | 76.53 | 80.22 | 74.97 |
| | Abduction/Adduction | 39.90 | 41.63 | 43.14 | 43.77 | 40.59 |
| | Rotation | 32.53 | 38.79 | 35.03 | 40.11 | 35.87 |
| Knee | Flexion/Extension | 76.98 | 77.31 | 69.96 | 77.55 | 69.85 |
| | Abduction/Adduction | 26.50 | 22.67 | 24.12 | 26.72 | 27.58 |
| | Rotation | 42.07 | 36.61 | 34.09 | 31.58 | 38.99 |
| Ankle | Flexion/Extension | 96.07 | 90.51 | 89.76 | 91.12 | 107.05 |
| | Abduction/Adduction | 50.51 | 45.24 | 43.30 | 46.63 | 49.08 |
| | Rotation | 62.15 | 51.79 | 65.50 | 57.27 | 59.34 |

**Table 4. Average ROM in degrees for kicking leg during right grand battement to the front during various degrees of warm-up (pre warm-up, 50%, 75%, 100%, and pointe)**

| | | Pre Warm-Up (°) | 50% (°) | 75% (°) | 100% (°) | Pointe (°) |
|---|---|---|---|---|---|---|
| Hip | Flexion/Extension | 103.55 | 108.78 | 103.340 | 106.96 | 102.65 |
| | Abduction/Adduction | 29.28 | 35.13 | 39.51 | 38.58 | 30.13 |
| | Rotation | 19.45 | 24.78 | 29.71 | 26.49 | 28.22 |
| Knee | Flexion/Extension | 58.94 | 60.41 | 97.45 | 69.55 | 67.18 |
| | Abduction/Adduction | 20.72 | 29.44 | 21.19 | 24.17 | 26.04 |
| | Rotation | 34.30 | 39.51 | 49.21 | 52.98 | 58.99 |
| Ankle | Flexion/Extension | 89.86 | 91.58 | 90.74 | 93.36 | 113.29 |
| | Abduction/Adduction | 70.52 | 60.60 | 57.54 | 72.95 | 59.43 |
| | Rotation | 71.89 | 56.13 | 46.45 | 81.57 | 65.72 |

Note. 75% warm-up ROM average does not include Subject B.

**Table 5. Average ROM in degrees for kicking leg during right grand battement to the side during various degrees of warm-up (pre warm-up, 50%, 75%, 100%, and pointe)**

| | | Pre Warm-Up (°) | 50% (°) | 75% (°) | 100% (°) | Pointe (°) |
|---|---|---|---|---|---|---|
| Hip | Flexion/Extension | 107.18 | 116.27 | 119.03 | 119.70 | 111.82 |
| | Abduction/Adduction | 78.98 | 79.43 | 81.95 | 80.36 | 80.17 |
| | Rotation | 48.30 | 55.46 | 60.89 | 62.14 | 58.39 |
| Knee | Flexion/Extension | 56.46 | 58.94 | 62.79 | 63.47 | 67.74 |
| | Abduction/Adduction | 27.69 | 20.08 | 19.90 | 24.80 | 27.23 |
| | Rotation | 44.08 | 40.84 | 43.44 | 57.54 | 50.37 |
| Ankle | Flexion/Extension | 90.18 | 88.01 | 95.23 | 100.45 | 115.34 |
| | Abduction/Adduction | 58.15 | 43.24 | 64.63 | 74.99 | 99.62 |
| | Rotation | 56.53 | 67.33 | 60.46 | 68.85 | 65.19 |

**Table 6. Average ROM in degrees for kicking leg during right grand battement to the back during various degrees of warm-up (pre warm-up, 50%, 75%, 100%, and pointe)**

| | | Pre Warm-Up (°) | 50% (°) | 75% (°) | 100% (°) | Pointe (°) |
|---|---|---|---|---|---|---|
| Hip | Flexion/Extension | 85.07 | 66.59 | 76.66 | 65.99 | 71.85 |
| | Abduction/Adduction | 37.36 | 38.37 | 40.14 | 40.64 | 37.65 |
| | Rotation | 34.77 | 32.64 | 40.21 | 42.20 | 39.39 |
| Knee | Flexion/Extension | 69.82 | 40.33 | 75.77 | 57.39 | 69.13 |
| | Abduction/Adduction | 28.02 | 21.81 | 35.30 | 21.62 | 25.80 |
| | Rotation | 39.11 | 31.39 | 39.68 | 33.77 | 35.95 |
| Ankle | Flexion/Extension | 90.55 | 87.83 | 98.69 | 91.72 | 119.78 |
| | Abduction/Adduction | 69.08 | 52.48 | 69.84 | 59.78 | 64.52 |
| | Rotation | 61.13 | 41.47 | 47.71 | 50.06 | 66.45 |

**Table 7. Average jump height in millimeters for sautés during various degrees of warm-up (50%, 75%, and 100%)**

| | 50% (mm) | 75% (mm) | 100% (mm) |
|---|---|---|---|
| Average Jump Height | 305.83 | 313.35 | 328.35 |

**Table 8. Average ROM in degrees for sautés during various degrees of warm-up (50%, 75%, and 100%)**

| | 50% (°) | | 75% (°) | | | 100% (°) | |
|---|---|---|---|---|---|---|---|
| | Left | Right | Left | | Right | Left | Right |
| Hip | Flexion/Extension | 54.10 | 53.28 | 52.42 | 51.13 | 54.97 | 53.29 |
| | Abduction/Adduction | 33.40 | 28.46 | 31.00 | 30.51 | 32.89 | 34.18 |
| | Rotation | 19.85 | 16.55 | 23.39 | 16.34 | 26.51 | 20.55 |
| Knee | Flexion/Extension | 100.00 | 102.92 | 95.43 | 98.10 | 101.49 | 105.65 |
| | Abduction/Adduction | 22.11 | 23.93 | 35.82 | 24.54 | 27.67 | 27.42 |
| | Rotation | 58.19 | 48.40 | 51.55 | 48.59 | 51.91 | 52.27 |
| Ankle | Flexion/Extension | 111.45 | 111.05 | 113.48 | 118.95 | 111.70 | 120.61 |
| | Abduction/Adduction | 42.28 | 65.30 | 56.76 | 72.52 | 52.40 | 71.39 |
| | Rotation | 63.05 | 56.97 | 66.14 | 65.91 | 65.95 | 62.08 |

**Table 9. Average flexion/extension for the hip, knee, and ankle in degrees for takeoff leg (left leg) during right grand jeté during various degrees of warm-up (50%, 75%, and 100%)**

| | 50% (°) | | | 75% (°) | | | 100% (°) | | |
|---|---|---|---|---|---|---|---|---|---|
| | Back Leg (left) | Front Leg (right) | Hip Angle at Peak of Jump | Back Leg (left) | Front Leg (right) | Hip Angle at Peak of Jump | Back Leg (left) | Front Leg (right) | Hip Angle at Peak of Jump |
| Hip Flexion/ Extension | -44.46 | 82.32 | 126.78 | -42.73 | 81.93 | 124.66 | -45.54 | 82.79 | 128.32 |
| Knee Flexion/ Extension | 14.20 | -6.25 | | 13.91 | -7.23 | | 17.54 | -3.72 | |
| Ankle Flexion/ Extension | -63.07 | -64.54 | | -66.86 | -66.87 | | -62.54 | -64.76 | |

**Table 10. Average jump height in millimetres for right grand jeté during various degrees of warm-up (50%, 75%, and 100%)**

| | 50% (mm) | 75% (mm) | 100% (mm) |
|---|---|---|---|
| Average Jump Height | 316.41 | 314.17 | 350.20 |

## Kinetics – EMG

The following figures show the effect that warm-up had on the left and right quadricep, hamstring, and gastrocnemius muscles.

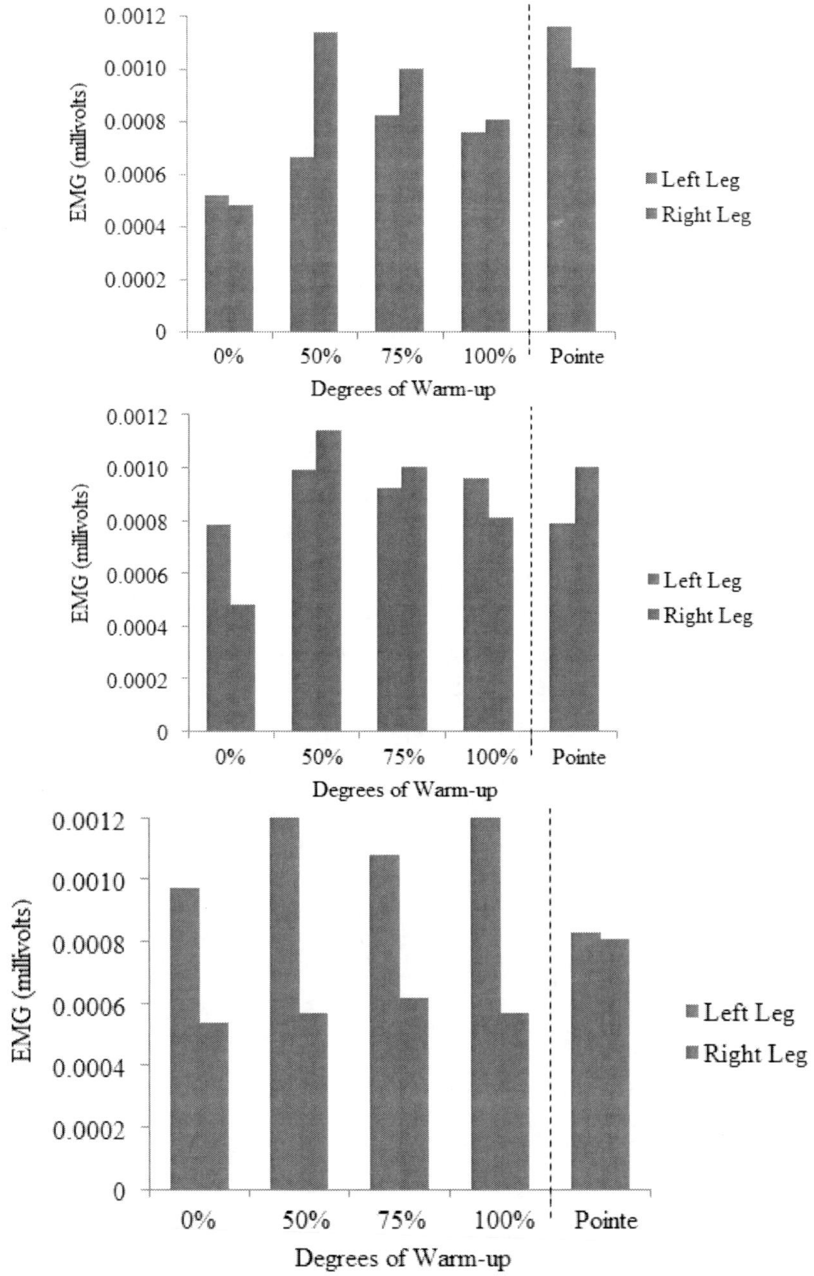


Note. 75% warm-up EMG average does not include Subject B.

Figure 4. Average peak EMG for the quadricep (top), hamstring (middle), and gastrocnemius (bottom) muscles of the kicking leg during left (blue) and right (red) grand battement to the front at various degrees of warm-up (0%, 50%, 75%, 100%, and pointe).

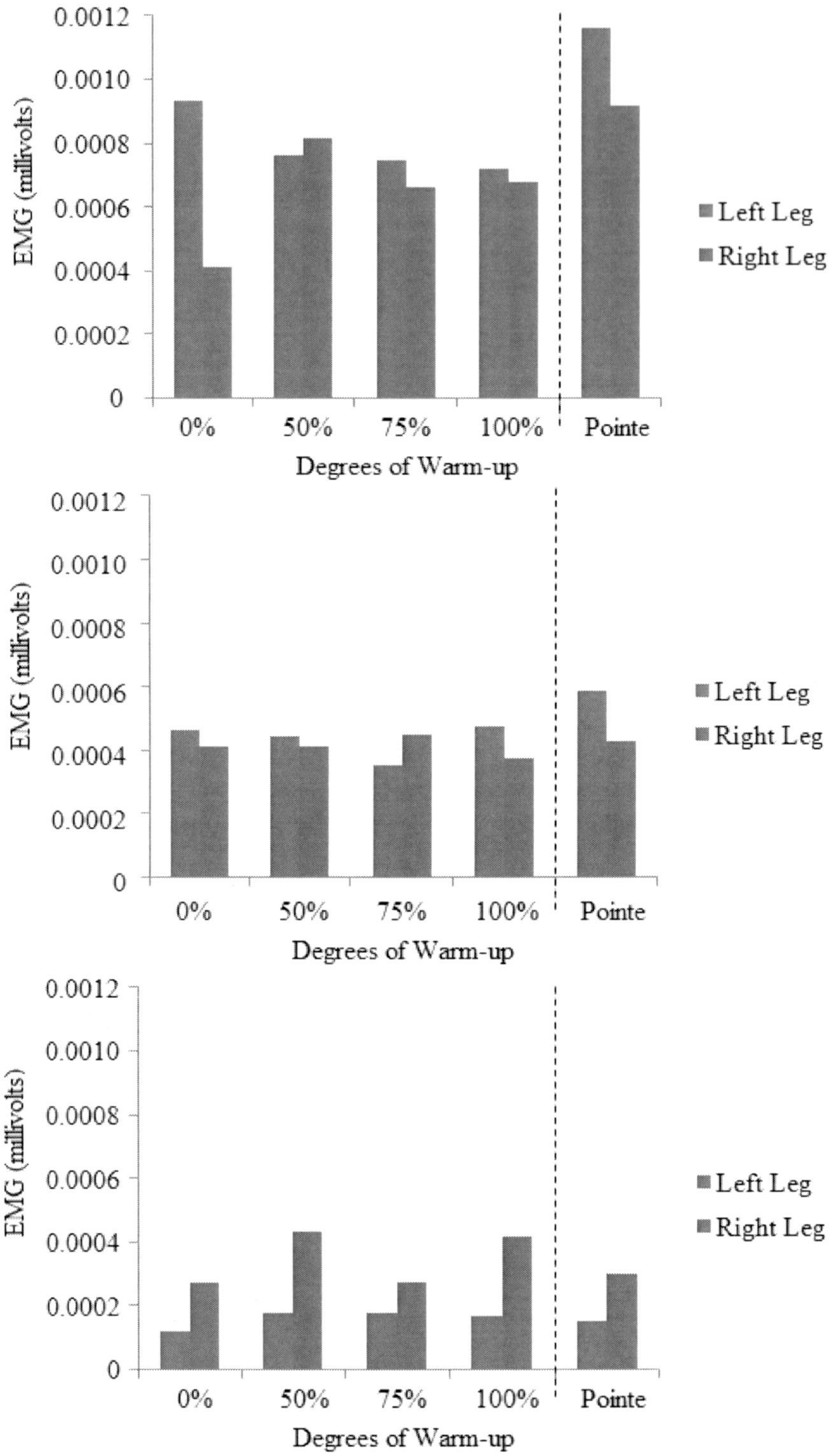


Figure 5. Average peak EMG for the quadricep (top), hamstring (middle), and gastrocnemius (bottom) muscles of the kicking leg during left (blue) and right (red) grand battement to the side at various degrees of warm-up (0%, 50%, 75%, 100%, and pointe).

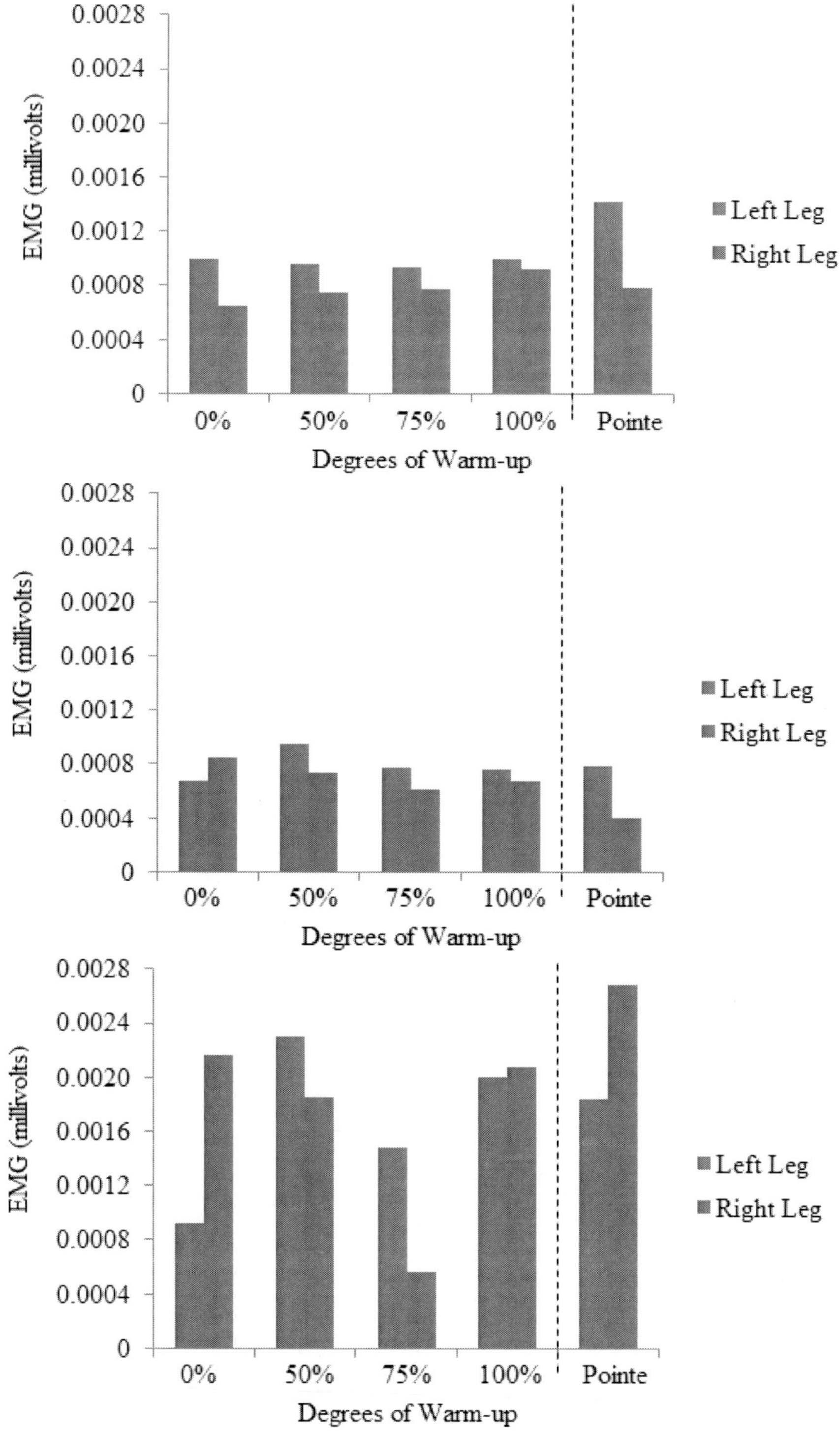


Figure 6. Average peak EMG for the quadricep (top), hamstring (middle), and gastrocnemius (bottom) muscles of the kicking leg during left (blue) and right (red) grand battement to the back at various degrees of warm-up (0%, 50%, 75%, 100%, and pointe).

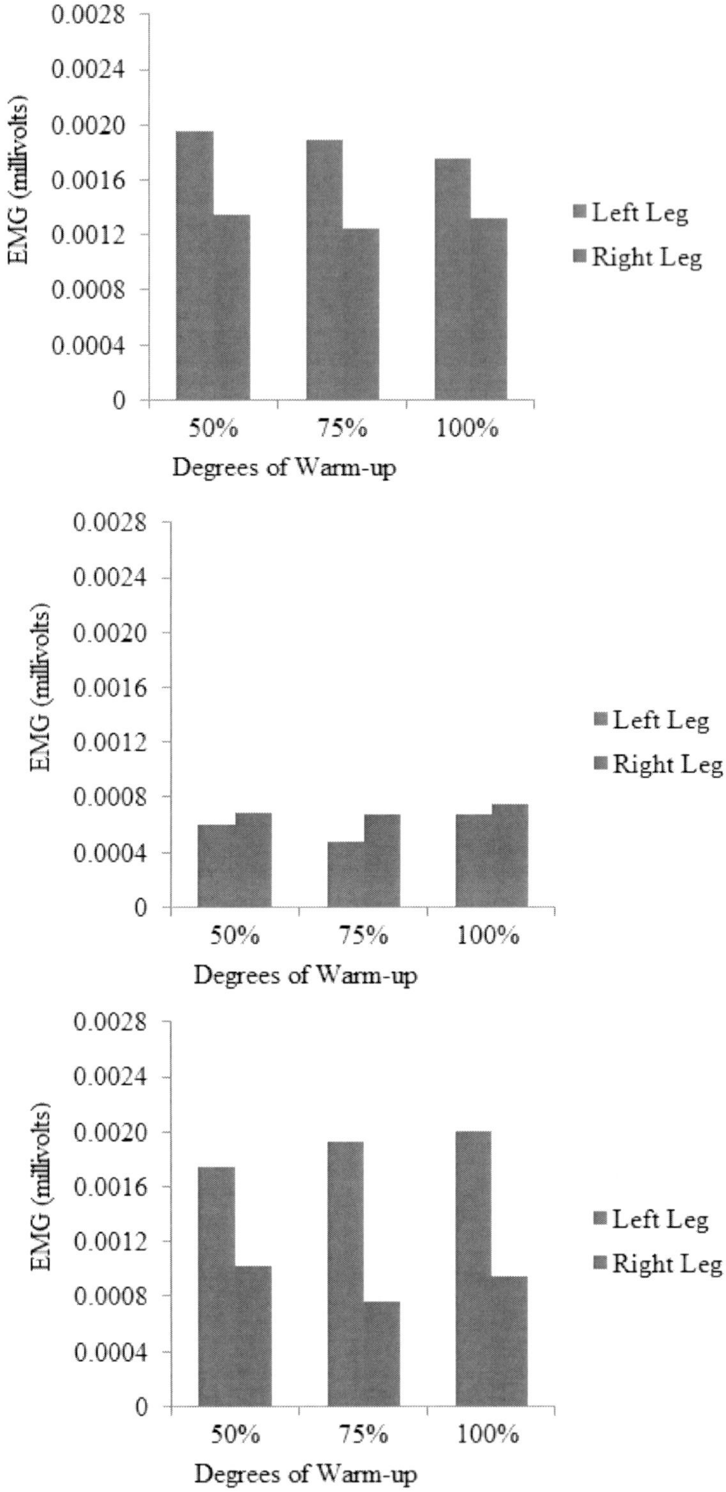


Figure 7. Average peak EMG for the quadricep (top), hamstring (middle), and gastrocnemius (bottom) muscles of the left (blue) and right (red) legs during sautés at various degrees of warm-up (50%, 75%, and 100%,).

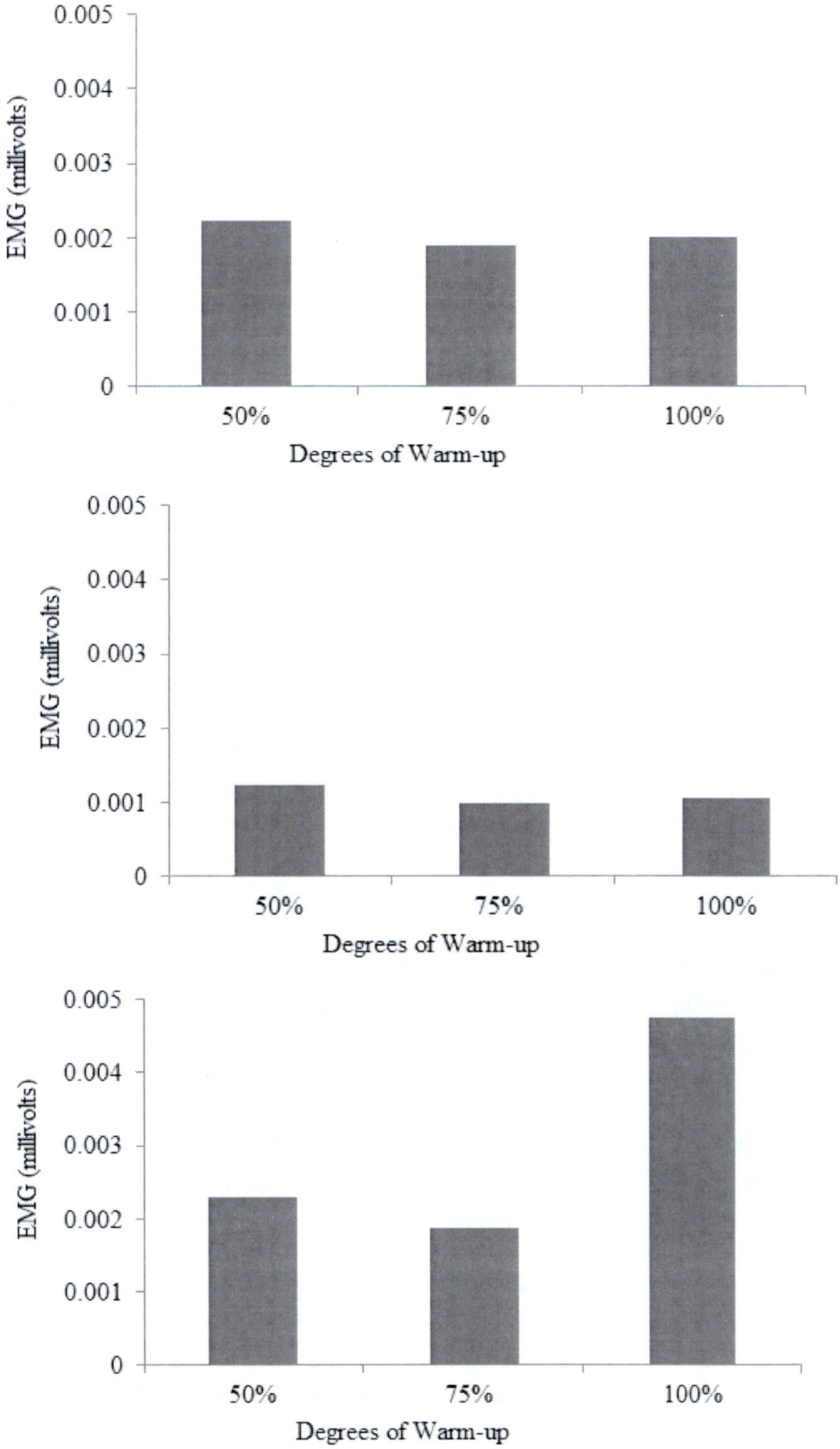


Figure 8. Average peak EMG for the quadricep (top), hamstring (middle), and gastrocnemius (bottom) muscles for back leg (left) during grand jeté at various degrees of warm-up (50%, 75%, and 100%,).

# DISCUSSION

This study examined the effects that warm-up has on risk of injury in ballet dancers. The results revealed that as the dancer becomes more warmed-up, they generally have increased flexibility or ROM and decreased EMG or muscle activity. The hypothesis that was initially proposed, suggesting that warm-up can decrease a dancer's risk of becoming injured has yet to be been verified but, if muscle activity and ranges of motion can be used as indicators of stress, it seems reasonable to pursue longitudinal studies that may confirm the consequential nature of warm-up routines.

Findings for the small sample of dancers of the current study revealed that each joint, specifically the hip, knee, and ankle joints, reaches peak ROM at different degrees of warm-up. It is also evident that there is some asymmetry between the dancers' right and left joints during certain techniques. Furthermore, the results show that there is some asymmetry in the EMG values of the right and left legs. Each of these phenomena provide food for thought for further studies.

Different joints reach peak ROM at different stages of warm-up. On average, the hip joint displayed peak flexion/extension at 50% warm-up. Abduction/adduction was not significantly influenced by warm-up. But, trends do show that during grand battement to the front, as warm-up increased, the dancers kicked their leg more towards the midline of their body (i.e., navel). This is actually what the traditions of the discipline require dancers to aim for. Ideally. they should try to align their kicking foot with the insole of their supporting foot, which should be in line with their navel if their weight is over their supporting leg. Moreover, on average, the hip joint reached peak rotation at 75%-100% warm-up. Therefore, the hip joint tends to reach peak ROM at about 75% warm-up. The knee also seems to reach peak ROM at about the same time. Results show that on average the knee joint reached peak flexion/extension at 75%-100% warm-up. Peak abduction/adduction of the knee joint was typically reached at 50% warm-up. Rotation of the knee joint was reached at 100% warm-up.

Interestingly, the ankle always reached peak flexion/extension during pointe work. This could be because the ankle is in plantarflexion (extension) to support the weight of the dancer during the grand battement or it could be because the ankle joint actually takes a lot longer to warm-up when compared to the hip and knee joints. EMG data also corresponds with this finding. The gastrocnemius muscles typically showed the highest activity during pointe work (Figure 4; Figure 5; Figure 6). This is probably because the dancer is standing on pointe and the gastrocnemius muscles are continuously engaged. As a result, the gastrocnemius muscles experience considerable strain during pointe work. Also, on average, peak abduction/adduction of the ankle joint was reached at 0% warm-up. Peak rotation of the left ankle was reached at 50% warm-up, whereas, peak rotation of the right ankle was reached at 100% warm-up.

When dancers warm-up prior to ballet class they tend to begin by warming-up their hamstring muscles and hip joints with various stretching exercises. The results from this study suggest that ballet dancers should also focus on warming-up their knee joints, but more importantly, they should spend extra time warming-up their ankle joints. As previously discussed, some aspects of ROM of the ankle joint warm-up quickly (abduction/adduction), whereas other aspects (flexion/extension) may take longer than other joints. This is important when considering injury prevention. In a study by Conti and Stone (1995) it was found that approximately 10%-30% of musculoskeletal injuries occurring in ballet are associated with the ankle joint. This is important for dancers to know because during demi-pointe (going up onto the tips of the toes) and pointe work, ballerinas force their ankle into extreme plantarflexion and this can contribute to many common ankle injuries suffered by ballet dancers. Because flexion/extension of the ankle takes so much longer to warm-up, dancers should focus on warming-up their ankles more during preliminary stretches and exercises.

Furthermore, sauté results show some asymmetry between the left and right legs. A sauté jump is a small vertical jump that should use both legs equally. Therefore, it seems logical that both legs would have similar results for flexion/extension, abduction/adduction, and rotation for the hip, knee, and ankle joints. However, this tends to not be the case. Abduction/adduction of the ankles varied by 15°-23°, with the right ankle consistently abducting more (Table 8). This could be another cause of the high rate of ankle injury. Previous research has found that the ankle motion during a relevé on pointe is slightly different for the dominant leg versus the nondominant leg (Cheng-Feng, Fong-Chin, & Hong-Wen, 2005). A relevé on pointe is very similar to a sauté jump (but, the toes remain in contact with the floor during a relevé). Therefore, there could be differences in ankle ROM between the dominant and nondominant legs during a sauté. Both of the dancers that participated in this study appeared to have a stronger and more dominant right leg, a fact that could be the cause of the asymmetry between the two ankle joints.

Average jump height was calculated for both the sauté and the grand jeté. Jump height increased as the dancer became more warmed-up. Peak height of the sauté was reached at 100% warm-up (Table 7). For the sauté, hip, knee, and ankle joints typically had the greatest ROM at 100% warm-up. Average jump height for the grand jeté is shown in Table 10. The grand jeté also reached peak jump height at 100% warm-up. At the peak jump height, the largest hip angle occurred. Had the dancers in this study been of professional level, a hip angle of ~180° would have been expected. However, because the subjects were high-level amateurs, their hip angles were less than 180°.

Regarding EMG, on average, muscle activity for the hamstring and quadricep muscles seemed to slightly decrease as the dancers became more warmed-up (Figure 8). Therefore, it may be inferred that the efficiency of muscle use in the left leg (leg used for takeoff) improved, because jumps were higher and muscle activity was lower as warm-up progressed. On the other hand, activity of the gastrocnemius muscle of the takeoff leg seemed to increase at 100% warm-up, indicating them to be under more strain. The larger effort of the gastrocnemius could be the main cause of the increase in jump height. If the dancer pushes off the ground with more force, then it would be expected that she jump higher.

For grand battement, EMG results showed the activity of the right leg muscles to be relatively consistent (~0.0006Mv), whereas, there is more variation in the activity of the left leg muscles (Figure 4; Figure 5; Figure 6). Therefore, for the dancers that participated in this study, the left leg muscles were under more strain than the right leg muscles. This could potentially be due to differences in strength between the dominant and non-dominant legs.

Another interesting result found in this study is evident when correlating EMG data with ROM data. During left grand battement to the front at 50% warm-up, the hamstring muscles reached peak EMG values (~0.0010mV) when the left hip joint reached peak flexion/extension (Figure 4). During grand battement, dancers are taught to "'lift from under the leg' or [to] 'use hamstrings to lift the leg, not the quads'" in order to "achieve greater height of the leg with less 'effort'" (Clippinger, 2007, p. 209). It seems as though the dancers were able to execute this approach to technique during the grand battement exercise. However, as the dancers became more warmed-up, and perhaps more fatigued, the hamstring muscles decreased in activity as the quadriceps muscles took over. If this phenomenon was indeed because the hamstring muscles fatigued, it could advantageous for dancers to focus less on the hamstring muscles during warm-up so that they do not fatigue so quickly during the class proper.

This study, as well as similar research from Malliou et al. (2007) and McGuinness and Doody (2006), provides important information for ballet instructors to consider when developing a program that considers injury prevention. As McGuinness and Doody (2006) discovered, it may be possible to decrease dancers' injury rate by at least 10% with appropriate warm-up. It is critical that dancers and choreographers know that different joints reach full ROM at different degrees of warm-up because the hip joint may not take very long to reach full ROM, but the ankle joint takes a lot longer. Also, sauté and grand jeté display peak jumping height at 100% warm-up. Therefore, in order for dancers to reach full height of their grand allegro (big jumps) during a performance, they should warm-up to almost 100%.

An important limitation of this study is that the sample size was very small. As such it represents a comparative case study only. In order to generalize these results, the protocol of the current study would need to be used in conjunction with a larger sample size. As well, the determination of the different degrees of warm-up was somewhat subjective. Warm-up stage was emperically determined by the ballet instructor. A more objective determination of stage of warm-up may not actually be possible since it is influenced by environmental conditions (temperature, type of garments worn, etc.) and the subjects perceptions.

Despite the above-mentioned limitations, the findings of this study point toward important issues for ballet dancers. Understanding the effect and sequencing of warm up on ROM, muscle activation, is essential for injury prevention. The findings of this study has potential application for the general public as well. An awareness that warm up is differential across muscle groups and joint flexibilities suggests that warm up activities should be tailored to the demands of the specific exercise.

Finally, some areas that could be investigated in future research include: 1) looking at the effect of warm-up on injury rate among professional ballet dancers, 2) an exploration of length of warm-up needed to optimize injury prevention, 3) design of specific stretches and exercises to reduce injury rates, and 4) looking at the effects of cool-down on injury rate in ballet dancers.

## CONCLUSION

The current study underpins the importance of warm up in preparation for vigorous physical activity. To decrease potential variability among subjects, this research examined two highly trained ballet dancers of approximately the same skill level. Results demonstrated warm up to be an essential component of the activity as joint ranges of motion increased during the course of warm up while muscle activity decreased. From this, it is reasonable to conclude that preliminary exercise is important to prepare dancers' bodies for ballet classes and performances. A thoughtful warm up routine may consequently be a useful tool for dance instructors and choreographers to help reduce injury rates among their dancers. As well the conclusions drawn with regard to dancers may be taken as a prescription for reducing injury rates for vigorous activity of any kind.

## REFERENCES

Adam, M. U., Brassington, G. S., Steiner, H., & Matheson, G. O. (2004). Psychological factors associated with performance-limiting injuries in professional ballet dancers. *Journal of Dance Medicine & Science*, 8 (2), 43-46.

Cheng-Feng, L., Fong-Chin, S., & Hong-Wen, W. (2005). Ankle biomechanics of ballet dancers in relevé en pointé dance. *Research in Sports Medicine*, 13(1), 23-35.

___(2007). Chapter 4: The pelvic girdle and hip joint. In K. S. Clippinger, *Dance Anatomy and Kinesiology: Principles and Exercises for Improving Technique and Avoiding Common Injuries* (pp. 157-227). Champaign, IL: Human Kinetics.

___ (2005). Before Class. In E. Gaynor Minden, *A Ballet Companion: A Dancer's Guide to the Technique, Traditions, and Joys of Ballet* (pp. 107-120). New York: Simon & Schuster, Inc.

Haight, H. J. (1998). Morphologic, physiologic, and functional interactions in elite female ballet dancers. *Medical Problems of Performing Artists*, 13(1), 4.

___ (2007). Ch 3: Kinetic concepts for analyzing human motion. In S. J. Hall, *Basic Biomechanics* (5th ed.) (pp. 61-80). New York: McGraw-Hill Companies, Inc.

___ (2007). Ch 5: The biomechanics of human skeletal articulations. In S. J. Hall, *Basic Biomechanics* (5th ed.) (pp. 119-138). New York: McGraw-Hill Companies, Inc.

Kennedy, J. G., Hodgkins, C. W., Colombier, J.-A., Guyette, S., & Hamilton, W. G. (2007). Foot and ankle injuries in dancers. *International SportMed Journal*, 8(3), 141-165.

Mara, T. (1987). *The Language of Ballet: A Dictionary*. Princeton, NJ: Princeton Book Company Publishers.

___ (2007). Warm-up (preliminary exercise). In W. D. McArdle, F. I. Katch, & V. L. Katch, *Exercise Physiology Energy, Nutrition, & Human Performance* (6th ed.) (pp. 588-590). Philadelphia: Lippincott Williams & Wilkins.

McGuinness, D., & Doody, C. (2006). The injuries of competitive Irish dancers. *Journal of Dance Medicine & Science,* 10(1/2), 35-39.

Shah, S. (2008). Caring for the dancer: Special considerations for the performer and troupe. *Current Sports Medicine Reports*, 7(3), 128-132.

Shan, G. (2005). Comparison of repetitive movements between ballet dancers and martial artists: Risk assessment of muscle overuse injuries and prevention strategies. *Research in Sports Medicine*, 13(1), 63-76.

Shan, G., & Bohn, C. (2003). Anthropometrical data and coefficients of regression related to gender and race. *Applięd Ergonomics*, 34(4), 327-337.

Shan, G., & Westerhoff, P. (2005). Full-body kinematic characteristics of the maximal instep soccer kick by male soccer players and parameters related to kick quality. *Sports Biomechanics,* 4(1): 59-72.

Twitchett, E., Brodrick, A., Nevill, A. M., Koutedakis, Y., Angioi, M., & Wyon, M. (2010). Does physical fitness affect injury occurrence and time loss due to injury in elite vocational ballet students? *Journal of Dance Medicine & Science*, 14(1), 26-31.

Visentin, P. & Shan, G. (2011). Applications of EMG Pertaining to Music Performance - A Review. *Arts BioMechanics*. 1(1):15-32.

Winter, D. A. (1990). Biomechanics and Motor Control of Human Movement. New York: Wiley.

In: Advances in Arts Biomechanics
Editors: Gongbing Shan and Peter Visentin
ISBN: 978-1-53615-641-6


*Chapter 10*

# LANDING KINEMATICS DURING THE GRAND JETÉ CONSIDERING PEAK LANDING FORCE

***Yasuyuki Yoshida*[1], *Ai Matsuura*[2] *and Mayumi Kuno-Mizumura*[2]**
[1]Tokyo Institute of Technology, Tokyo, Japan
[2]Ochanomizu University, Tokyo, Japan

## ABSTRACT

The grand-jeté is a fundamental split leap movement in classical ballet that requires single-leg landing. This study was designed to investigate the magnitude differences in the landing force through lower limb kinematics during the grand-jeté. An eight-camera motion analysis system was used to obtain three-dimensional lower extremity kinematic data of 14 female classical ballet dancers. One force plate was used to measure the ground reaction forces. Participants performed grand-jetés and landed on the force plate. The pelvis, thigh, leg, and foot were derived for kinematic analysis. The participants were divided into the following two groups on the basis of the normalized peak ground reaction force in the vertical direction: the HIGH group and the LOW group. The jumping distance was greater in the HIGH group than in the LOW group during the flight phase ($p < 0.05$). The LOW group had a significantly greater hip flexion angle at touchdown on the force plate during the landing phase ($p < 0.05$). Furthermore, the duration of knee flexion was greater in the LOW group ($p < 0.05$). These findings may provide potential information for classical ballet dancers and teachers that the characteristic changes of the movements might be able to reduce the magnitude of impact loading to ensure safe landings in the grand-jeté.

**Keywords:** classical ballet, landing, kinematics, ground reaction forces

## INTRODUCTION

The grand-jeté, a fundamental split leap movement in classical ballet, requires single-leg landing. Because ballet technique shoes and pointe shoes do not have shock-absorbing materials, classical ballet dancers must use lower limb movement strategies to control landing.

Few reports in the literature describe studies examining split leap movements in classical ballet. In ballet leaps and jumps, landing has a greater risk for injury to the soft tissues of the knee joint compared to take-off (Kulig, Fietzer, and Popovich. 2011) Dancers with patellar tendinopathy have been shown to have greater vertical and braking landing impulses than healthy dancers (Fietzer, Chang, and Kulig, 2012). Some studies have assessed the movements made by rhythmic gymnasts during split leaps (Di Cagno, Baldari, Battaglia, Monteiro, Pappalardo, Piazza, and Guidetti, 2009; Dyhre-Poulsen, 1987). These earlier studies did not investigate the magnitude of differences in peak landing forces by considering lower limb kinematics between injury-free dancers during the split leap movements.

There is one report about the landing movements during split leaps performed by injury-free dancers. According to Wyon, Harris, Brown, and Clark (2013), a significant negative correlation was found between the peak knee flexion angle and the peak force during landing in the grand-jeté performed by injury-free dancers. Only the peak flexion angle for the knee joint was investigated. The peak flexion angle of the hip and ankle joints and other kinematic parameters, such as the angle at touchdown on the floor and the duration from touchdown to the peak angle, were not investigated. Classical ballet dancers might improve their landing performance without sustaining injury if the change in the lower limb movement could reduce landing forces.

This study provides an increased understanding for performing the grand-jeté. A previous study has shown that the foot and ankle region is vulnerable to injury in classical ballet dancers (Nilsson, Leanderson, Wykman, and Strender, 2001). Excessive impact loading may have negative effects on the body, resulting in injuries (Zhang, Derrick, Evans, and Yu, 2008). Therefore, it would be beneficial for dance teachers and dancers to know the movements that could reduce impact loading to ensure safe landings in the grand-jeté.

This study was designed to investigate the magnitude differences in the landing force through lower limb kinematics during the grand-jeté, a split leap movement, in classical ballet.

# METHODS

## Participants

This study examined 14 female classical ballet dancers, including professional dancers and dance students. The mean age was 24.1 ± 5.4 years, the mean height was 1.6 ± 0.1 m, and the mean mass was 47.6 ± 4.6 kg. The mean training was 19.4 ± 5.7 years, and the mean age at starting ballet was 4.8 ± 1.9 years. No participant had any orthopedic or neurological condition that might have affected the trials. Before participating, all participants provided written consent after being informed of the possible risks and benefits. The University Institutional Biomedical Research Ethics Committee granted research approval.

## Experimental Setup

To obtain three-dimensional lower extremity kinematic data, an eight-camera Vicon motion-analysis system (MX; Vicon Motion Systems, UK) with a sampling rate of 250 Hz was used. Twenty-three reflective markers were placed on the greater trochanter, lateral thigh, medial epicondyle, lateral epicondyle, lateral shank, medial malleolus, lateral malleolus, toe, and heel on both sides, and the right anterior superior iliac spine, left anterior superior iliac spine, right llium, left llium, and sacrum. To measure the ground reaction force (GRF), a Kistler force plate (9286BA; Kistler Instruments AG, Switzerland) with a sampling rate of 1000 Hz was used. The force plate was time-synchronized to the motion analysis system. A right-handed system was used for defining directions. The positive Z-axis represented an upward vertical direction. The travelling direction of the grand-jeté was the positive Y-axis. The positive X-axis represented the right direction for participants.

## Protocol

All participants wore their own, regularly used classical ballet technique shoes and performed grand-jetés until they landed on the force plate thrice. The participants started performing grand-jetés 4 m away from the force plate. They took a short running approach to take off using the left leg, landed on the force plate on the right leg, and then ran away from the force plate. During the flight phase, the left arm was extended in the forward direction, with the arm positioned in the second arabesque position, and the right arm was extended and aligned with the dancer's shoulder. During dynamic data collection, the participants were instructed to perform the grand-jeté as usual.

## Data Analysis

The landing phase was defined as the moment between which the foot first contacted the force plate and the moment it last contacted the force plate. The three-dimensional kinematic and the GRF data were processed using Kwon3D motion analysis software (Version XP; Visol Inc., Korea). The marker coordinate data were filtered using a 7-Hz low-pass Butterworth filter. The local coordinate systems for the pelvis, thigh, leg, and foot were defined from the markers, allowing for six degrees of freedom for each segment. The X-, Y-, and Z-axes of the segmental frames were aligned with the mediolateral, anteroposterior, and longitudinal axes of the segments, respectively. The flexion angles for the hip and knee and the plantarflexion and dorsiflexion angles for the ankle in the landing leg, that is the right leg, were extracted. If the ankle angle was more than 90 degree, it was considered dorsiflexion, and if it was less than 90 degree, it was plantarflexion.

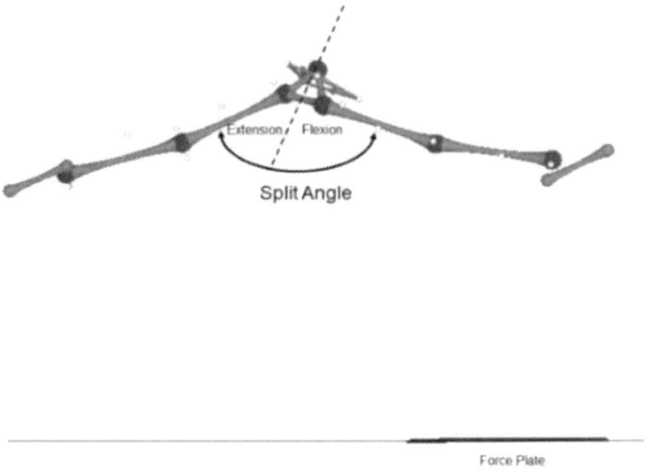


Figure 1. Definition of the split angle.

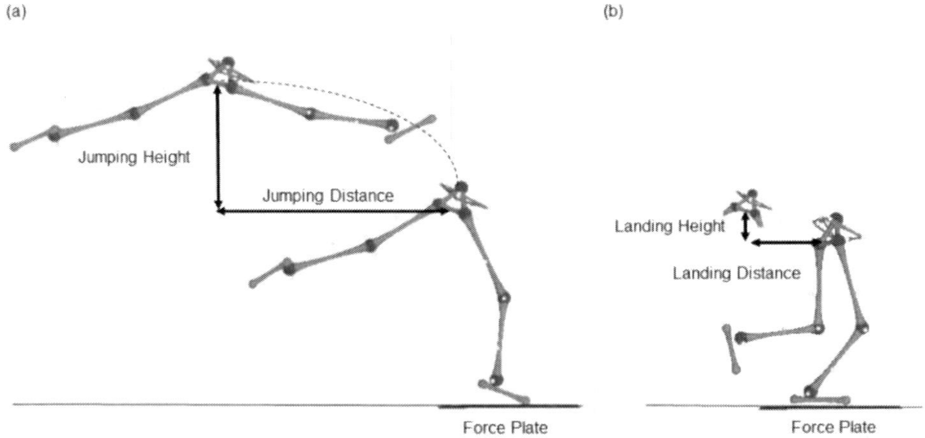


Figure 2. Definitions of spatial characteristics: (a) jumping distance and height, (b) landing distance and height.

Figure 1 shows the definition of a split angle. During the flight phase in the grand-jeté, the participants were required to split the hip joint between flexion for the right hip and extension for the left hip. The peak value was determined as the summation of the absolute value of the flexion angle in the right hip and the extension angle in the left hip. Figure 2 shows the definitions of spatial characteristics. The jumping distance and the jumping height were calculated using position data for the midpoint of the right and left hip joints. The values for the peak height and touchdown on the force plate were determined. These coordinate differences were calculated in the sagittal plane. The landing distance and the landing height were also calculated. The midpoint of the lowest position of the hip joints was determined on the force plate. Coordinate differences between the lowest position and the position at touchdown on the force plate were calculated during the landing phase. These distances and heights were normalized considering the participants' body height.

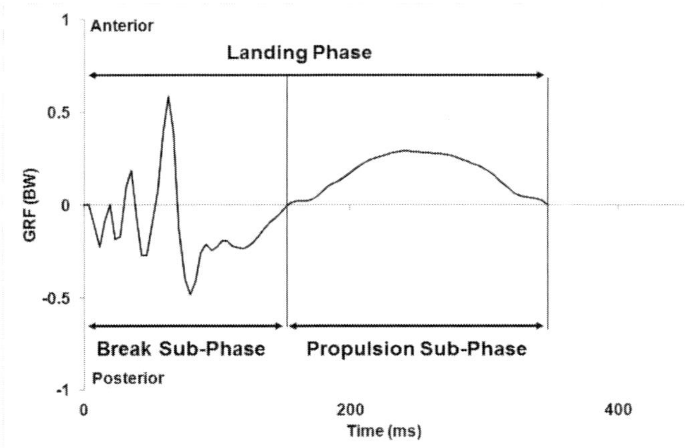


Figure 3. Definitions of the break sub-phase and the propulsion sub-phase on the basis of the anterior–posterior ground reaction force.

The GRF data were normalized considering each participant's body weight in the vertical and the anterior–posterior directions. Toe-on and toe-off were identified using a threshold of 20 N of GRF. The landing phase of the GRF was further divided into the break sub-phase and the propulsion sub-phase (Figure 3). On the basis of the anterior–posterior GRF, the GRF for the propulsion sub-phase was regarded as the GRF during anterior movement of the participants on the floor, corresponding to approximately the second half of the movement. The remainder such as first half was set as the GRF for the break sub-phase.

## Grouping and Statistical Analysis

For all participants, the mean of the three trials was calculated for each variable. The 14 participants were divided into the following two groups on the basis of the normalized peak GRF in the vertical direction: the HIGH group and the LOW group. The peak GRF was arranged in the descending order. The values in the top half were considered the HIGH group, and the values in the other half were considered the LOW group.

All statistical procedures were conducted using SPSS software (Version 21.0; IBM Corp., USA). The means and standard deviations were calculated for each kinematic and GRF variable during the flight phase and the landing phase. For each dependent variable, two-tailed independent samples t-tests were used to evaluate group differences at a significance level of $p < 0.05$.

# RESULTS

## Group Characteristics

All participants were divided into two groups based on the peak GRF in the vertical direction. The mean ± standard deviation of the peak vertical GRF was 3.29 ± 0.41 BW. The characteristics of the HIGH and the LOW groups are listed in Table 1.

**Table 1. Descriptive characteristics (mean ± SD) of participants in the HIGH and LOW groups**

| Characteristics | HIGH | LOW |
|---|---|---|
| Peak Vertical GRF (BW) | 3.60 ± 0.28 | 2.98 ± 0.24 |
| Age (years) | 22.3 ± 3.1 | 26.0 ± 6.7 |
| Body Height (m) | 1.6 ± 0.1 | 1.6 ± 0.1 |
| Body Mass (kg) | 47.9 ± 5.6 | 47.3 ± 3.8 |
| Ballet Experience (years) | 17.9 ± 3.8 | 20.9 ± 7.1 |
| Starting Age (years) | 4.4 ± 1.9 | 5.1 ± 2.0 |

## GRF

Figure 4 shows the time curves of the GRF in the vertical and anterior–posterior directions in all trials for both the groups during the landing phase. The statistical comparisons of all variables considering GRF are listed in Table 2. The peak vertical GRF occurred significantly earlier in the HIGH group ($p < 0.05$).

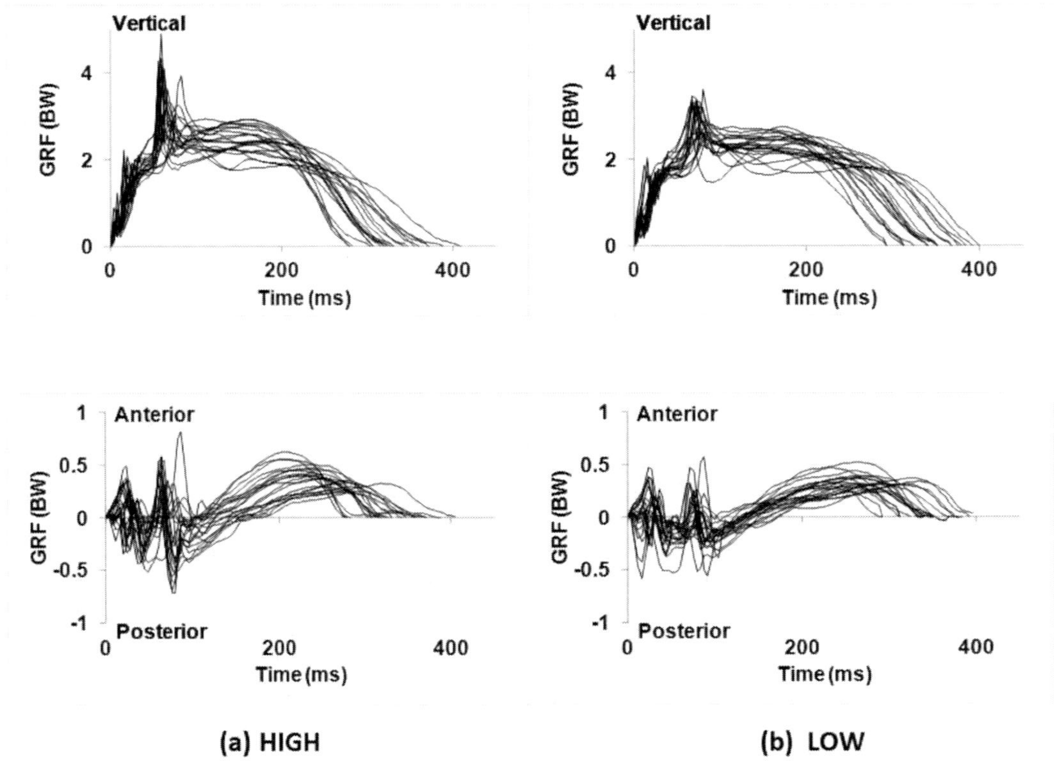


Figure 4. Ground reaction force curves in the vertical direction (upper) and in the anterior–posterior direction (below) for all trials during landing in the grand-jeté: (a) HIGH group, (b) LOW group.

**Table 2. Ground reaction force parameters (mean ± SD) in the HIGH and LOW groups during landing in the grand-jeté, $p < 0.05$**

| | HIGH | | | LOW | | | |
|---|---|---|---|---|---|---|---|
| Variable | Mean | ± | SD | Mean | ± | SD | |
| Landing Phase Duration (ms) | 324.57 | ± | 31.18 | 338.67 | ± | 26.30 | |
| Break Sub-Phase Duration (ms) | 127.05 | ± | 17.38 | 133.90 | ± | 14.11 | |
| Break Sub-Phase Duration (%) | 39.05 | ± | 2.30 | 39.68 | ± | 4.62 | |
| Peak Fymin (BW) in break sub-phase | -0.36 | ± | 0.17 | -0.30 | ± | 0.09 | |
| Peak Fymax (BW) in break sub-phase | 0.37 | ± | 0.13 | 0.25 | ± | 0.14 | |
| Peak Fymax (BW) in propulsion sub-phase | 0.39 | ± | 0.09 | 0.34 | ± | 0.06 | |
| Time to Peak Fymin (ms) in break sub-phase | 67.24 | ± | 10.61 | 53.71 | ± | 28.43 | |
| Time to Peak Fymax (ms) in break sub-phase | 44.76 | ± | 10.15 | 42.86 | ± | 14.20 | |
| Time to Peak Fymax (ms) in propulsion sub-phase | 244.38 | ± | 26.35 | 266.29 | ± | 28.21 | |
| Time to Peak Fzmax (ms) | 64.00 | ± | 9.27 | 72.38 | ± | 3.90 | * |

## Kinematics

The statistical comparisons of all variables about kinematics are listed in Table 3. The jumping distance during the flight phase was greater in the HIGH group than in the LOW group ($p < 0.05$). No significant difference was found in the peak split angle. The LOW group had a significantly greater hip flexion angle at touchdown on the force plate during the landing phase ($p < 0.05$). Furthermore, the duration of knee flexion was greater in the LOW group ($p < 0.05$).

**Table 3. Lower limb kinematic parameters (mean ± SD) during landing in the grand-jeté TD: Touchdown on the force plate $p < 0.05$***

| | | HIGH | | | LOW | | | |
|---|---|---|---|---|---|---|---|---|
| Phase | Parameters | Mean | ± | SD | Mean | ± | SD | |
| Flight | Peak Split Angle (deg.) | 132.96 | ± | 11.72 | 119.95 | ± | 12.77 | |
| | Jumping Distance (%Height) | 29.92 | ± | 2.02 | 25.59 | ± | 3.52 | * |
| | Jumping Height (%Height) | 10.30 | ± | 1.21 | 9.54 | ± | 1.76 | |
| Landing | Hip Flexion at TD (deg.) | 46.27 | ± | 6.34 | 54.05 | ± | 6.27 | * |
| | Knee Flexion at TD (deg.) | -20.96 | ± | 3.28 | -21.27 | ± | 6.39 | |
| | Ankle Plantarflexion at TD (deg.) | 63.69 | ± | 2.04 | 63.92 | ± | 1.92 | |
| | Peak Hip Flexion (deg.) | 48.26 | ± | 6.60 | 55.55 | ± | 7.32 | |
| | Peak Knee Flexion (deg.) | -45.95 | ± | 4.84 | -47.47 | ± | 5.00 | |
| | Peak Ankle Dorsiflexion (deg.) | 128.81 | ± | 3.49 | 126.81 | ± | 4.96 | |
| | ΔHip Angle (deg.) | 1.89 | ± | 2.32 | 1.40 | ± | 1.91 | |
| | ΔKnee Angle (deg.) | 24.99 | ± | 5.60 | 26.20 | ± | 6.17 | |
| | ΔAnkle Angle (deg.) | 65.11 | ± | 4.27 | 62.89 | ± | 4.89 | |
| | Duration of Hip Flexion (ms) | 53.52 | ± | 47.50 | 48.00 | ± | 50.06 | |
| | Duration of Knee Flexion (ms) | 144.57 | ± | 9.36 | 160.19 | ± | 12.25 | * |
| | Duration of Ankle Dorsiflexion (ms) | 186.29 | ± | 14.29 | 199.43 | ± | 14.86 | |
| | Landing Distance (%Height) | 19.23 | ± | 1.39 | 18.37 | ± | 2.28 | |
| | Landing Height (%Height) | 11.18 | ± | 0.96 | 11.53 | ± | 1.23 | |

## DISCUSSION

The peak GRF in the vertical direction was 3.60 ± 0.28 BW and 2.98 ± 0.24 BW for the HIGH and LOW groups, respectively. The landing force in split leap movements has been reported previously. According to Fietzer et al. (2012), the peak GRF in the vertical direction during landing in the saut de chat, a leap similar to a grand-jeté, was 4.37 ± 0.82 BW and 5.95 ± 1.11 BW for the control group and patellar tendinopathy group, respectively. Wyon et al. (2013) who investigated the landing force during three continuous grand-jetés reported the maximum values to range from 3.88 ± 0.54 BW to 4.29 ± 1.02 BW. The different values might be owing to the difference in instructions for the experiment. Previous studies instructed participants to jump at their full performance level. In contrast, the participants in this study were instructed to move as usual, similar to the instructions given in the study by Dyhre-Poulsen (1987). Consequently, the peak value in our study differed greatly from those found in previous studies.

One of the important results of this study is that the time of appearance of the peak vertical GRF was approximately 8 ms earlier in the HIGH group during the landing phase; this was significantly different between the groups. Furthermore, the peak vertical GRF was approximately 21% higher in the HIGH group than in the LOW group. Surprisingly, other parameters about GRF were not significantly different between the two groups. GRF peaks are commonly used to evaluate the landing technique (Niu, Wang, He, Fan, and Zhao, 2011). The GRF is the product of the mass of the dancer and deceleration on the floor (Hackney, Brummel, Jungblut, and Edge, 2011). Therefore, the LOW group may have a better landing technique than the HIGH group.

The duration of knee flexion during the landing phase was significantly different between the HIGH and the LOW groups. According to Devita and Skelly (1992), the duration of contact with the floor from initial contact to peak knee flexion was longer for soft landing than that for stiff landing. Wyon et al. (2013) reported a negative correlation between the peak knee flexion angle and the peak force during landing in the grand-jeté. In the present study, only the duration of knee flexion, not the peak knee flexion angle, was different between the groups. As mentioned before, the time of appearance of the peak vertical GRF was significantly different. Therefore, the duration of knee flexion might be correlated with reducing the peak GRF and delaying the time of appearance. Although not reported in the previous study, the time of appearance and the duration would probably be different.

Furthermore, the hip flexion angle at touchdown during the landing phase was significantly different between the groups. The hip joint flexion angle after touchdown varied only slightly in the grand-jeté movement. Blackburn and Padua (2009) found that trunk flexion reduced the landing forces and the quadriceps activity during drop landings. It is therefore indicated that hip joint movement might not contribute to impact absorption in the grand-jeté. However, the initial angle of the hip joint at touchdown might affect quadriceps muscle activity, because the rectus femoris muscle in the quadriceps is a biarticular muscle. Knee extensors have an important role in impact force attenuation during landing activities (Decker, Torry, Wyland, Sterett, and Richard Steadman, 2003). Landing performance during the landing phase is controlled by eccentric contractions of lower extremity extensors (Wu, Zhang, Liu, Zhang, and Xie, 2013). Although knee extensor muscle activities were not available for our investigation, the larger hip flexion angle at touchdown combined with knee flexion may be one of the important factor for reducing landing forces.

Interestingly, the peak split angle was not significantly different between the groups. Before landing in the grand-jeté, the task of splitting legs is demanding. A greater GRF might not require a greater change in the split angle. During the flight phase, the jumping distance was longer in the HIGH group than in the LOW group. However, the jumping height was not significantly different between the groups. Numerous skill-related factors while executing the grand-jeté might influence performance. The technical leaping performance was probably more dependent on stiffness, floor reaction, coordination, and gymnastic skills rather than on strength (Di Cagno et al. 2009). However, in the current study, it is not clear if the landing distance in the HIGH group was influenced by the selection strategy.

A limitation of this study was that classical ballet dancers performed the grand-jeté in a laboratory condition, which is not the same as performing it on stage. Dancing on stage might be influenced by other cognitive demands, such as music, choreography, and space, which were not considered in the present study. Nevertheless, the findings of this study provide sufficient information about landing in the grand-jeté.

## Conclusion

Classical ballet dancers had a longer duration of knee joint flexion, a larger hip flexion angle at touchdown, and a shorter jumping distance, all of which reduces impact loading during landing in the grand-jeté. These findings may provide potential information for classical ballet dancers and teachers that the characteristic changes of movements might be able to reduce the magnitude of impact loading to ensure safe landing in the grand-jeté.

## Acknowledgment

This study received financial support from KAKEN (22700614).

## References

Blackburn, J. T., and Padua, D. A. (2009). Sagittal-plane trunk position, landing forces, and quadriceps electromyographic activity. *Journal of Athletic Training*, 44:174-179.

Decker, M. J., Torry, M. R., Wyland, D. J., Sterett, W. I., and Richard Steadman, J. (2003). Gender differences in lower extremity kinematics, kinetics and energy absorption during landing. *Clinical Biomechanics*, 18:662-669.

Devita, P., and Skelly, W. A. (1992). Effect of landing stiffness on joint kinetics and energetics in the lower extremity. *Medicine and Science in Sports and Exercise*, 24:108-115.

Di Cagno, A., Baldari, C., Battaglia, C., Monteiro, M. D., Pappalardo, A., Piazza, M., and Guidetti, L. (2009). Factors influencing performance of competitive and amateur rhythmic gymnastics--gender differences. *Journal of Science and Medicine in Sport*, 12:411-416.

Dyhre-Poulsen, P. (1987). An analysis of splits leaps and gymnastic skill by physiological recordings. *European Journal of Applied Physiology and Occupational Physiology*, 56:390-397.

Fietzer, A. L., Chang, Y. J., and Kulig, K. (2012). Dancers with patellar tendinopathy exhibit higher vertical and braking ground reaction forces during landing. *Journal of Sports Sciences*, 30:1157-1163.

Hackney, J., Brummel, S., Jungblut, K., and Edge, C. (2011). The effect of sprung (suspended) floors on leg stiffness during grand jeté landings in ballet. *Journal of Dance Medicine and Science*, 15:128-133.

Kulig, K., Fietzer, A. L., and Popovich, J. M. Jr (2011). Ground reaction forces and knee mechanics in the weight acceptance phase of a dance leap take-off and landing. *Journal of Sports Sciences*, 29:125-131.

Nilsson, C., Leanderson, J., Wykman, A., and Strender, L. E. (2001). The injury panorama in a Swedish professional ballet company. *Knee Surgery, Sports Traumatology, Arthroscopy*, 9:242-246.

Niu, W., Wang, Y., He, Y., Fan, Y., and Zhao, Q. (2011). Kinematics, kinetics, and electromyogram of ankle during drop landing: a comparison between dominant and non-dominant limb. *Human Movement Science*, 30:614-623.

Wu, X., Zhang, S., Liu, Y., Zhang, D., and Xie, B. (2013). Do knee concentric and eccentric strength and sagittal-plane knee joint biomechanics differ between jumpers and non-jumpers in landing? *Human Movement Science*, 32:1299-1309.

Wyon, M., Harris, J., Brown, D., and Clark, F. (2013). Bilateral differences in peak force, power, and maximum plié depth during multiple grande jetés. *Medical Problems of Performing Artist*, 28:28-32.

Zhang, S., Derrick, T. R., Evans, W., and Yu, Y. J. (2008). Shock and impact reduction in moderate and strenuous landing activities. *Sports Biomechanics*, 7:296-309.

# SECTION 3. VISUAL AND MEDIA ARTS

In: Advances in Arts Biomechanics
Editors: Gongbing Shan and Peter Visentin
ISBN: 978-1-53615-641-6


*Chapter 11*

# Realtime Performance Using Biomechanical Gesturing Mediated via Technological Interfaces to Empower an Individual Partial Control of an Aesthetic Experience

***A. William Smith***
University of Lethbridge, Lethbridge, Alberta, Canada

## Introduction

Have you ever thought of a future with the possibility of being able to project what you imagine directly in the time and space of our shared reality? This could be for others to experience or for the imagining individual to explore. For instance, if I were indoors in winter and imagined being surrounded by trees lush with green leaves, golden sunshine, blue sky breezes and wanted to swirl magenta flower-petal-like objects through the scene, petal swarms that seemed to follow me around, or have them drift here and there according to how I waved my arm...

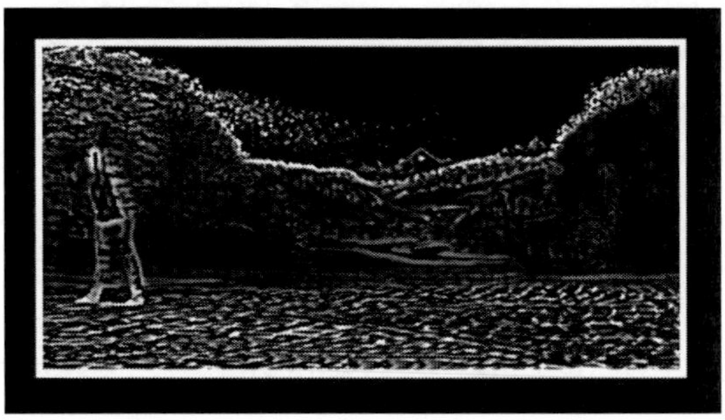

Figure 1. Mr & Mrs Dream—"Magic Box" Technological Backstage (Dassault Systemes, 2013)].

** For this figure, and nearly all others—especially those referring to website material—the author has provided freely interpreted aesthetic imagery that encapsulates the essence for him and suggests an experience a reader might have when accessing cited content.

. . .well this is nearly exactly what one sees occur by a performer and team in the Mr. & Mrs. Dream performance [see moment 5:32/6:16 sec in the referenced video, Figure 1] But in this possible future, we would not constrained with mere literal representation of environmental reality as is in the descriptive example above, it could be a transfiguration into symbolic or abstract 'language'.

Through human movements, events that are aesthetic or expressive can occur with little delay. In other words, there is a causal connection. A human moves her body using muscles and bones in any xyz dimension, whole body, a single body part, several body parts, or combinations, and the resulting motions can trigger events. A gesture or position, the speed of gesturing, and many other biomechanical aspects such as acceleration and complexity or density of the movements can initiate outside-the-body events. At the moment of this research paper, the individual causing the events, be they visual or audible, has gained some control of this at the beginning of the 21st-century and, based on current trends, the individual will be gaining greater control as the century progresses.

Gaining power of expression of our innermost selves, our imagination, our dreams and being able to control how that happens is part of the mind-age. We've seen how the ancients sought control of their environment through hunting, gathering, herding, and agriculture, controlling resources such as water in dams and channels. Then as still today, humans also had to control their defenses against other humans. Especially in the 1400s, there was a sense in Western civilization to map physical space leading to exploration and domination of land/territory. At the same time we were metaphorically extending our sight through the telescope and mapping the heavens, and extending our vision into the microcosm through use of the microscope. In the 1600s, the first part of the industrial age is identified where steam power at first extends the capabilities of our body, making possible the figurative muscle power of steamboats instead of rowing or steam trains instead of pulling. Wanting power to illuminate the night and control other processes, the electricity age was born in the late 1800s leading to innovation in information transmission, having control through telegraph, telephone, radio, and television. In the second half of the 1900s, through control of science, the computer and internet allowed individuals more democratic control of information and personal expression. While we explore our human body through DNA sequencing to gain control of lifespan, our brain activities through MRI and EKG to gain control of intelligence and memory—while we go deeper and deeper into the physical body—we are going deeper and deeper into our imagined worlds. This is all too obvious in the popularity of immersion in video games where players participate as characters within the worlds they see on screen, causing many thinkers to exclaim that the 21$^{st}$-century will be defined by games (Zimmerman et al. 2013; Sicart, 2015). Human beings appear hard-wired to want control of their physical, social, and other realities, including that of aesthetic expression. This paper speculates on the thought that a daydream may in the near future become observable as a reconstructed reality in realtime.

## KEYWORDS/CONCEPTS

- *Realtime*—the temporal state where events in one state-of-being are affecting events in another state-of-being within microseconds or similar short time-lag. This might include a person's body being a certain location within the xy floor plane that might affect the soundscape or visual at that particular moment, or, for instance gestures effecting visuals in another moment. The key idea is cause and effect, where the effect is observable nearly immediately.

- *Performance*—the awareness or consciousness of the individual affecting events through movement that there is a connection to another state-of-being. Therefore, there is intent. A person's walking movements on a busy downtown street would not be considered performance.
- *Using*—participatory.
- *Biomechanical Gesturing*—physical positioning of body or limbs in the xyz [z up] space with the so-called center of the floor [space] within a defined volume as 0,0,0 occurring through time. It is not merely position or shape of the limbs, but it can include forces and other scientifically measurable aspects of the physical body. The gerund form, gesturing, is deliberately chosen to indicate a time-based occurrence.
- *Mediated*—involves determining the nature of the biomechanical gesturing to associate or to cause other aesthetic events to occur. Essentially, it is the intellectual selection or determination of what biomechanical positioning or forces of a performer to track.
- *Technological Interfaces*—involves a means through hardware and software to determine that the physical state of a performer has changed from an earlier moment. This is often done with an input device such as a camera connected to a computer with a program to compute differences of body or body parts relative to the volume.
- *Empower*—act of enriching a human to be able to do more than a human is able to do under normal moments of reality without technological or another form of intervention.
- *Individual*—an autonomous biped, a human being with a sense of self who is able to move body or body parts in space.
- *Partial Control*—because of the specialized nature of creating or using technological interfaces, including programming biomechanical data gathering and processing, creating or generating visual or audio assets, and then being able to run a performance with biomechanical actions, an individual is considered part of a collaborative effort in contemporary instances. Moreover, there are limitations to the concept of being able to totally control the experience because the possibilities of aesthetic expression are preconditioned to the choices made within the available compendium of space and technology. If an individual were to be able to go anywhere and generate for others her mindscape (not those sounds and imagery from some set of sounds and imagery), that would be total control.
- *Aesthetic Experience*—while it is possible for an individual using biomechanical gesturing to be in one location and the state of being of visual or aural data to be in another, normally the latter occurs in the same place as the former. There is a connection between the individual creating motion or being observed and visual or aural data that occurs as a result of individual's actions. Because choices have been made in what visual or aural data is possible and how it occurs within the temporal sequence of biomechanical gesturing, the combination is considered human expression. One might claim movement is thought that is translated into elements, an

action that is not meant to accomplish an expedient goal, e.g., it does not produce food, it does not remove trash, or it does not grasp a real screwdriver.

## Human Motion Capture Background in Fine Arts Expressions

Of course there could be Motion Capture (MoCap) animations where the end is not aesthetics, such as a demonstration of astronauts in a space shuttle simulation. And, while there are many applications in fine arts using human motion capture data for other ends, such as establishing a biomechanical standard for evaluating classical ballet training (Yanagisawa et al. 2013) or generating 2-d movement scores readable to those who read Labanotation (Choensawat et al. 2015), this section introduces the reader to a brief background of some non-realtime uses of movement data in human expression/aesthetics.

A standalone animation using MoCap data involves several steps: 1) motion is performed and *simultaneously* captured, 2) the data is processed and cleaned up, 3) it is exported to a skeleton, 4) it is edited or assembled as a motion sequence, 5) it is then mapped on a character placed in a constructed environment, 6) rendered, and 7) assembled in post-production with sound. The end product can be narrative—in the service of telling a story or abstract. Of course this process cannot be considered realtime; it happens over hours, days, weeks, months, and sometimes years.

An example of such an aesthetic non-narrative, or abstract standalone MoCap work might be found by researchers wanting to explore movement on characters in environments impossible to experience, for instance dancers on a large rock sitting on a high railway bridge (Figure 2). In Smith's practice, except for a very few collaborative works, he performs the motion and creates the visuals, so there is more agency and control in his work than that of most other MoCap artists.

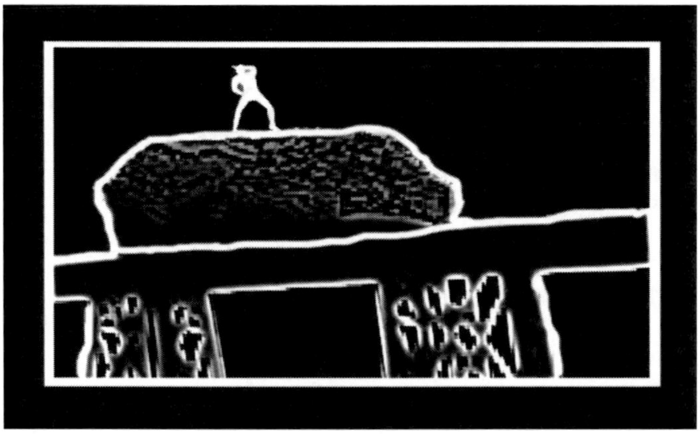

Figure 2. Example of a more traditional application of MoCap data (Smith, 2011).

Most individuals who create movement for aesthetic non-narrative standalones execute their motions and collaborators utilize their knowledge of the technologies to determine the visuals, as in the 1999 *Ghostcatching*, performed by Bill T. Jones (Figure 3). Here, as in the 1999 Merce Cunningham *Bipeds*, certain painterly textures, like brushstrokes, are applied to the surface of 3D model's skin and through the use of transparency of the skin/mesh. The result is a set of moving lines that one recognizes as human. By applying particle stream emitters with a timed decay, one can get smoke-like trails to follow selected points on the dancers' extremities and then disappear.

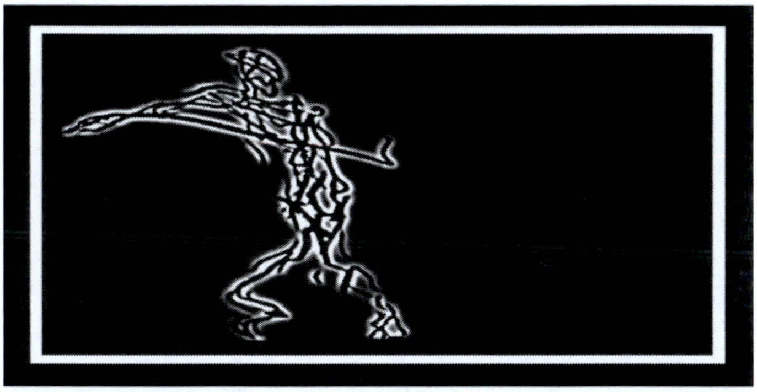

Figure 3. Example of artistic texturing of digital model (Canal de kinodanz, 2010).

Many examples of the narrative genre of standalone MoCap expressions are of high profile, such as *Tintin* (Jurgensen, 2011). They also engage what seems to be an army of collaborators and have a primary end of generating revenue.

As well, some time-based standalones use MoCap data for some characters that are composited into the scene, such as in the Hobbit ([N.A.], 2012). MoCap data can also be used to drive non-human characters, such as the dragon Smaug (9GAG.TV, 2016). In the full use of MoCap data as in *Tintin* or partial use as in the *Hobbit*, movement is created to serve an end, and is often representational.

Neither considered standalone in the case of pure MoCap animation nor composited with photoreality is that use of MoCap data in interactive systems where there is no input device of the player's biomechanical data. In gaming, MoCap data is often assembled on demand according to the moment of the story or location (e.g., by the pressure of certain buttons that might make a character to go forward and other buttons to turn, and so on), but there can be moments of standalone animation where interactivity is temporarily suspended. And, while many games involve MoCap data that are tests of skills for the players, some are focused on discovering how the story unfolds, such as in *Beyond: Two Souls* (Figure 4).

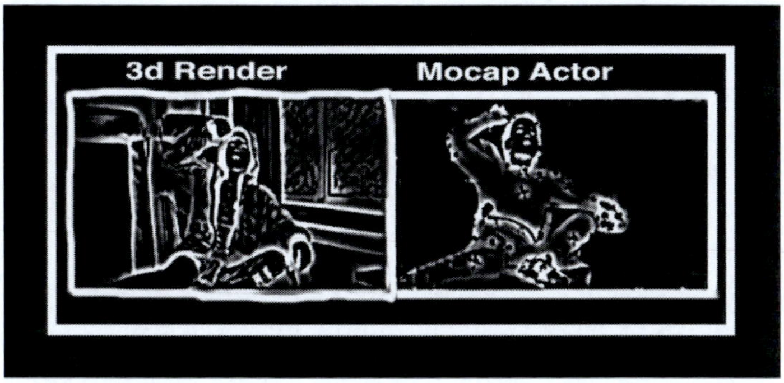


Figure 4. Example showing MoCap data being created and the resultant application in a game (GameSpot, 2012).

Aside from gaming being interactive and using MoCap data assembled on demand or as standalone, some interactive internet sites use MoCap data in the same ways. Many creators of MoCap content for the internet community Second Life, founded in 2003, advertise their services and products online (StudioMyANIMATION, 2011). There is an expansion however, where site visitors can customize or edit the motion through choices at an interface (Figure 5). For instance, a visitor to the site can choose certain positions for the arms (twenty offered here) or upper body actions and place them on a lower body stance of their character/avatar. While the intent of the visitor using this feature may be a desire to see herself embodied in a virtual character and execute certain actions as a means to explore the imagination, a researcher cannot dismiss that aesthetics might be involved in the choices that are made.

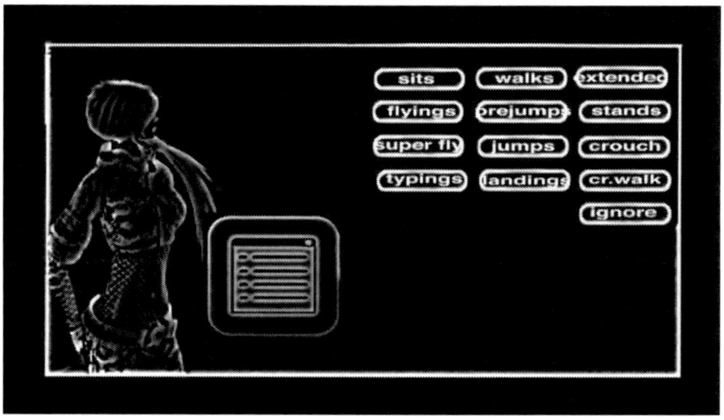


Figure 5. Visual interface for online modification of movement (VistaAnimation, 2010).

Storytelling itself within most taxonomies lies within the humanities, but when the storytelling leaves the realm of words written and spoken and become embodied or realized in a physical manner, aesthetics becomes important. In traditional media arts, the frame of the output device, the rectangular screen or monitor dominates many of the choices. There have been developments in the last decades exploring the sense of 3D that animations, films, and gaming have prepared the way for our contemporary cultures to be more attuned to illusions in real spaces.

## PSEUDO REALTIME PARTIAL CONTROL OF AESTHETIC EXPERIENCES

Through their biomechanical gesturing, some individuals appear in certain performance situations to be causing events, but in reality, they are not (Talbott, 2014). This perhaps reflects the desires of the collaborators involved to truly wish to see what such could be based on imagination (ENRA, 2013). Another example that might appear as if the performer controlled imagery, but really doesn't, begins with two trees growing, moving apart, joining, bearing an apple, then jumps into a street scene and continues evolving in a through-composed fashion, ending up with two performers with the illusion of large angel wings flapping (Figure 6).

This process for 'pseudo realtime partial control' might include the creation first of a standalone animation, and then set the positions of the bodies or movements of the performers to be at certain places at certain times for a performance. Or, it could be based first on a detailed time movement sequence then have visuals created and frontally projected, or superimposed. In any case within this 'pseudo' category, the individuals are following or matching the visuals, not causing or leading them.

An event with pseudo realtime partial control of aesthetic experiences can appear to an audience to be the same as an actual experience where individual agency is involved. This perhaps is a reflection that there is a contemporary will to link movement with aesthetic events, in this case, visual ones.

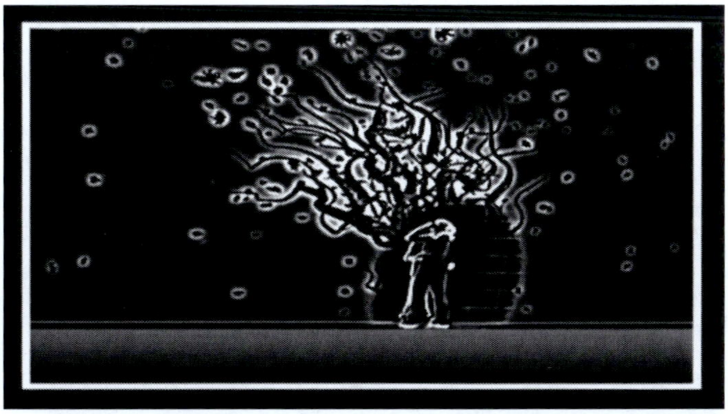

Figure 6. An example of human performers that are integrated with visuals with no realtime control (Kanal, 2011).

## REALTIME PARTIAL CONTROL OF AESTHETIC EXPERIENCES FOR AUDIENCES

### Realtime Mapping of a Costume Texture in a Real Environment

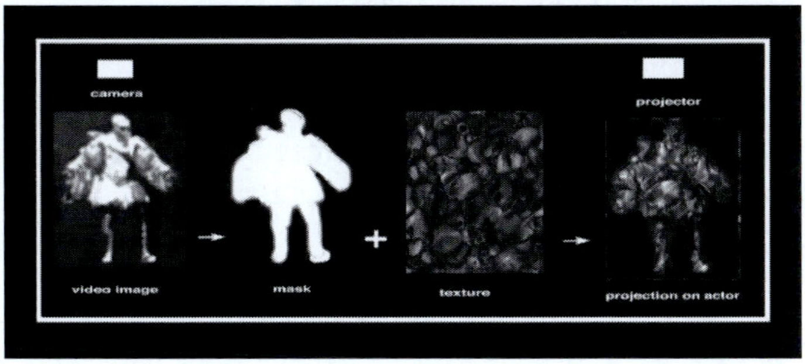


Figure 7. An example of using video-mapping of movement in order to create costume effects (Interaction, 2016).

In 2002 for the opera *The Jew of Malta*, movements of performers were tracked using a camera with realtime analyses of silhouettes (Figure 7). Textures were then projected on the performers as their 'virtual' costumes.

## Realtime Mapping to a Virtual Character/Avatar in a Virtual Environment

Another use of MoCap input is to drive a character onscreen, or, in other words, movements of a human performer is merely translated to a virtual avatar (Meador, 2004). In this work, live dancers were combined with a MoCap-driven dancing character at a 2004 event at Purdue, The performer driving the character had no control of computer graphics or other elements. Rather the screen was like a mirror, isolated and passive, where there was a one-to-one correspondence between the performer and her imagery. An exploration of technological challenges in this work seems to result in the subjugation of aesthetic goals to the process of collaboration. This is typical of the initial stages of collaborative art; often early adapters of technology must first overcome a fascination with the technology itself before moving on to sophisticated artistic applications.

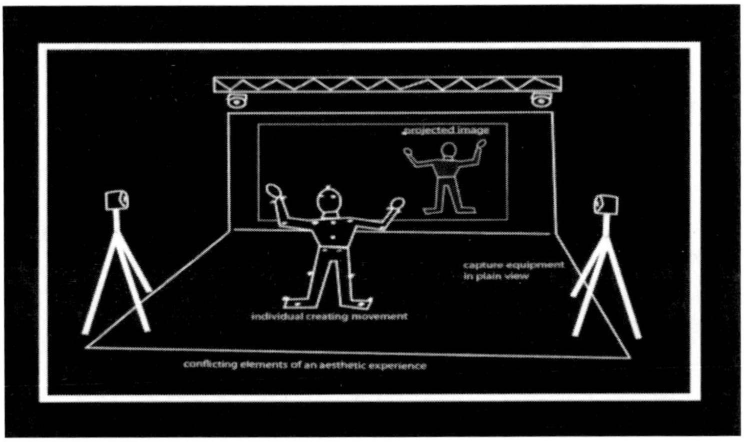


Figure 8. Illustration of how a traditional MoCap realtime performance could be aesthetically challenging in creating for an audience a suspended sense of disbelief.

An additional example of this form of translation of movement to imagery is found in the mapping to character/avatar in a virtual Shakespeare Rose Theatre which was projected on a single background plane (Delbigio, 2014). In Delbigio's study his Figure 1 depicts a character in Shakespearian costume projected as a reversed mirror image of the performer. His Figure 3 illustrates the performer facing outward with face to her right and a similar view of the avatar. The avatar is in a virtual set corresponding to a reconstruction of a Shakespeare Theatre. On stage left (or the audience right) the projection is split with the text of the performer must recite. In his Figure 4 we see a projection of the computer monitor with a far shot of the character facing stage right within a virtual theatre. A computer interface can be observed onscreen to the side while the performer faces stage left. The actor wears a MoCap suit with markers in the middle of the real space and one can see the MoCap cameras on either side of the screen. In his Figure 20 one sees a projection similar to his Figure 3 but one doesn't see the avatar. Rather, when the performer extends her hands in front of her, they appear pictured in the computer graphics. The shift in perspective is from 3rd person to 1st person. In his Figure 33 we see the hands in front of an actor facing downstage, but the projected imagery has hands directed upstage, into the virtual set of the Shakespeare Theatre. These figures of Delbigio illustrate some of the problems that can plague an aesthetic experience that employs MoCap. What is the relationship of the projected image to the real individual creating the motion? Does it face back, front, left, or right? As well, if the projected imagery is dwarfed by capture volume, there can be no illusion as the technology is all visible (including the markers on the MoCap suit the performer wears). And the performer is visible at the same time (Figure 8). These experiments appear self-referential.

## Realtime Mapping of Biomechanical Gesturing to Sonic Events

One of the earliest uses of motion to trigger aesthetic events utilized a video camera that had directed its feed to a grid within a software interface on a computer monitor (Rokerby, 1986). In essence this worked by examining how pixels from the background or unoccupied/empty space were modified by the presence of a body moving through the space.

Mark Coniglio developed a system called *midiDancer* in 1989 that had a wireless system measuring joint angles and allowed a performer to control sound. Coniglio and collaborator premiered work in 1994 as Troika Ranch.

More recently, a sampling of the activities at the Input Devices and Music Interaction Laboratory (IDMIL) can serve to illustrate part of the large body of work going on all over the world connecting gesture to sound (IDMIL-PM, 2016). One can see that there are various methods of connecting body positioning with, for instance, the angles of a clarinet to the performer's skeleton. Some, pictured, require optical systems as reflective markers adorn body parts.

During the capturing process, biomechanical gesturing can be seen as human-centric, requiring an active sense of the xyz of skeleton parts. Further methods of generating sound come from embedding sensors within the instruments themselves (IDMIL-PI, 2016). Those instruments requiring movement (movement includes orientation) serve as means to use motion to make an aesthetic experience. While musicians will rightly assert that all music requires their muscular action, the discussion here relates to realtime mediated gesturing via technological interfaces.

Using an affordable motion sensing technology such as the Kinect, new musical interfaces are being developed where a performer moves parts of her body in space without contacting a physical object (Dahl, 2014). Dahl concludes that for smooth movement controlling sound, the Kinect level of MoCap works okay, but it doesn't for precise, complex, rhythmic driven sounds.

Here it can be assumed that the movement-generating-sound performance experience can either be for audience consumption or for personal satisfaction, and will not be discussed in a later section focused on the self or solitary aesthetic experience.

## Realtime Mapping of Biomechanical Gesturing to Visuals and Sound Events

A project in 2005 used fifty markers distributed on four individuals with an optical MoCap system with sixteen cameras, driven by an AI program that had an associated gesture library (Mark, 2005). The gesture-capture algorithms could then look for patterns in the real-time data that matched the library and if there was a match, could trigger realtime output. One such example is when a certain pattern is found about the relationship of two dancers, branches develop from digital graphics that appear to reach out and connect the individuals. Other patterns cause other graphic movement and morphing on multiple screens. Four artists contributed to the image and sound palettes that held assets to create the experience with movement itself.

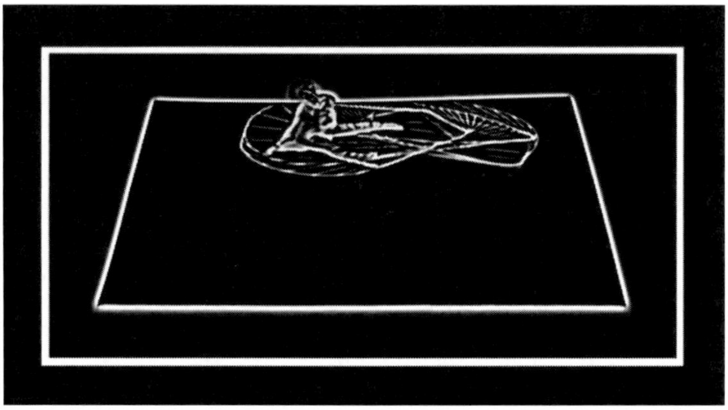

Figure 9. Projected imagery that is driven by movements of a performer (Biennale, 2016).

Frieder Weiss invented *Kalipso*, a video-motion-sensing software program and used it for the creation of realtime imagery of the 2006 solo dance *Glow* (Figure 9). One can see the imagery at several sites including (StageNoise, 2007) and the full version (Cultureband, 2014).

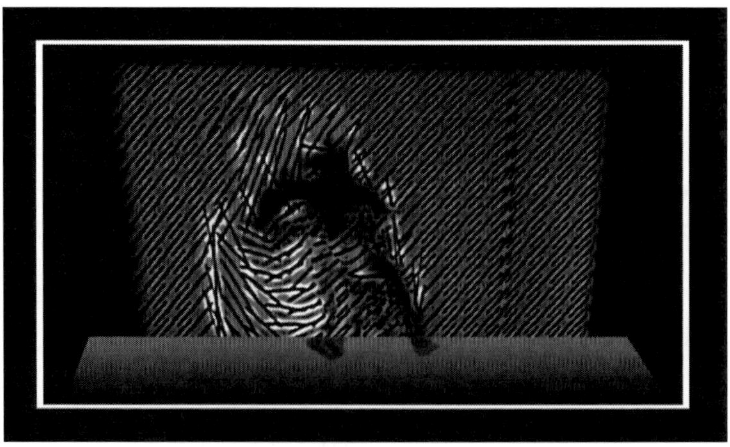

Figure 10. Example of projected visuals illustrating variety in aesthetic approach and limitations of performance space and movement (Weiss, 2010).

A similar approach to graphics using multiple performers can be seen in *Mortal Engine* (Figure 10). It mainly uses projection on the floor as well as what appears at moments to be frontal projection and top-down rear projection. The movement seems floor-bound and constrained, which could be a limitation of the technology.

Mark Coniglio developed *Isadora*, first released in 2002 that allowed for realtime processing of video based on realtime input. Notable in 2006 was a performance *16 [R]evolutions* (dancetechtv, 2012) and the analyses of body shape in terms of four parameters: straight, curved, complex, and size (Figure 11). The hardware is described as using an infrared camera and, even earlier than Frieder Weiss's system, predates the Xbox Kinect infrared input device by several years. With *Isadora*, the performers could interactively manipulate video and sound through their actions.

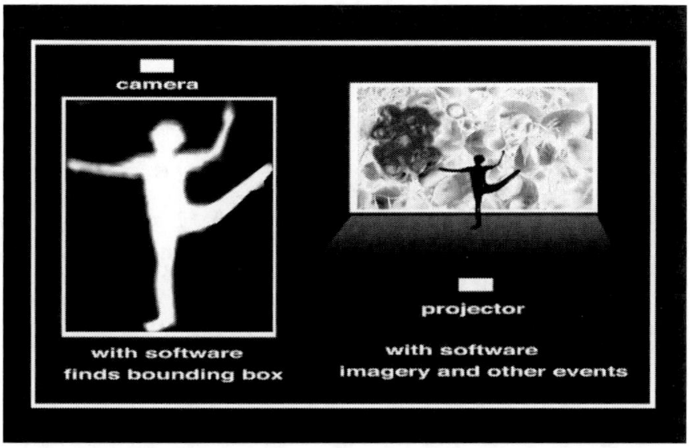


Figure 11. Diagram of the field-of-view and how size of movement is determined by a pioneer of allowing a performer to have some control of an associated aesthetic event (troikaranch, 2007).

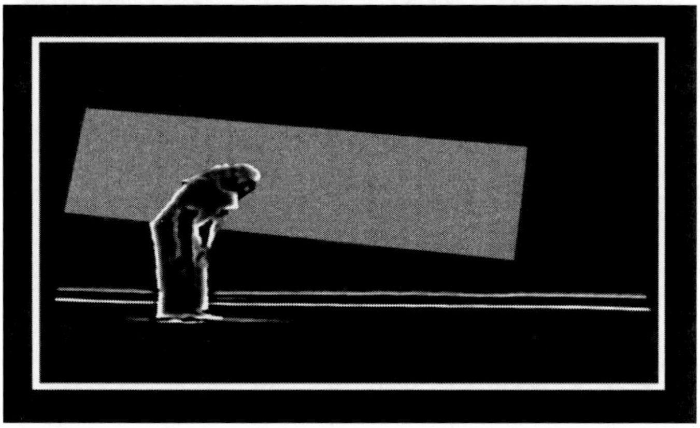

Figure 12. Example of experimenting with new technologies based on earlier platforms (Griffiths, 2007).

Incorporating *Isadora*, the project Videmote created a moving projection spot to display scenic imagery simultaneously dealing with keystoning and rotational variables of a moving video image (Figure 12). So, this was an advancement in options of how to output visuals to biomechanical gesturing, and serves to illustrate the continual process in technology where sophistication is built up on earlier platforms/achievements.

Using a camera to track where individuals were onstage and making associations for projecting imagery, Johnathan Bachrach facilitated Snappy Dance Theater's 2007 *String Beans* (Figure 13). The realtime-generated graphics with lines drawn at 15 frames a second were based on each individual's movements, and the effect was dubbed 'the scribbler'. The concept of realtime visuals as a mediated version of the humanoid performer is evident here, wherein there is a shift from object in the foreground to the background in which the object resides.

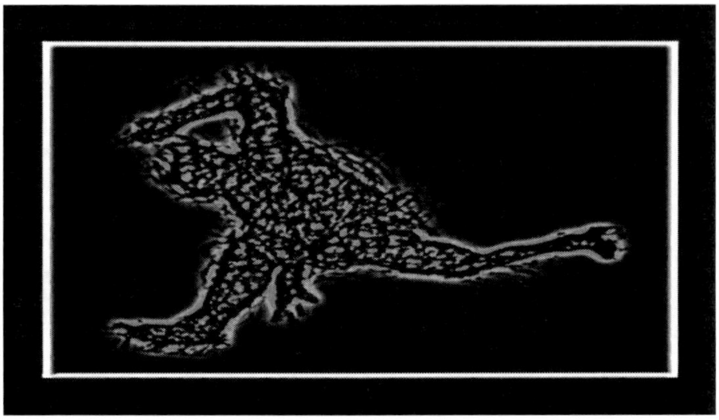

Figure 13. Example of the use of motion-tracking technology to create realtime projected versions of performers (Macel, 2010).

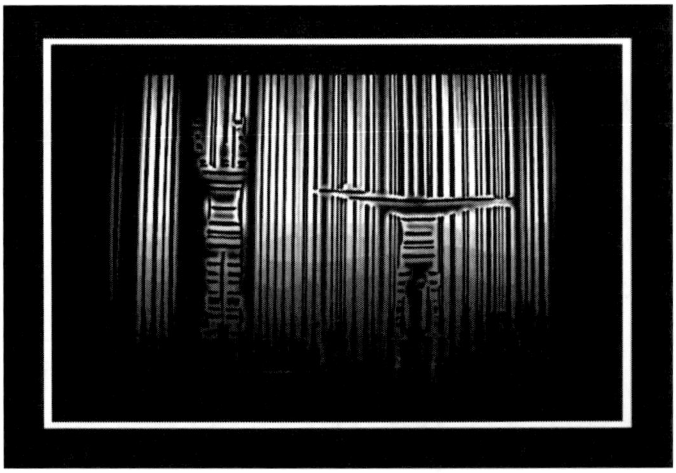

Figure 14. Example where movement parameters of performers affect the visual experience for an audience (Viennaexile, 2010).

In the 2010 *Apparition*, camera tracking was also employed, analyzing the moving outline/shape, speed, direction, intensity, and volume (deLahunta, 2013) created by the human performer. The visuals may have been influenced by movement but they were not controllable by the performer, suggesting that the sequence of visuals was somewhat predetermined. At one point in the work, the verticals in the background against the horizontals on the individuals in the foreground make the space especially kinematic (Figure 14).

Because creators of movements do not like to be constrained by any aspect of an inside-out or outside-in motion capture system and, as well, because those systems require use of markers, straps, wireless packs, or other hardware that must be worn in plain sight of an audience, we can see in hindsight that performing arts explorations using a videocam or infrared cam have been historically favored. It is no coincidence that the home videogame MoCap sensor xBox 360 Kinect released in 2010 became a viable and popular interface to connect motion with aesthetic effects (for details see Kinect, 2016). This system made MoCap technology affordable. Earlier systems were expensive and complex to operate, creating a power/knowledge/financial differential that privileged the artistic explorations only for select few individuals. With the xBox 360 Kinect, a democratization of artistic possibilities occurred.

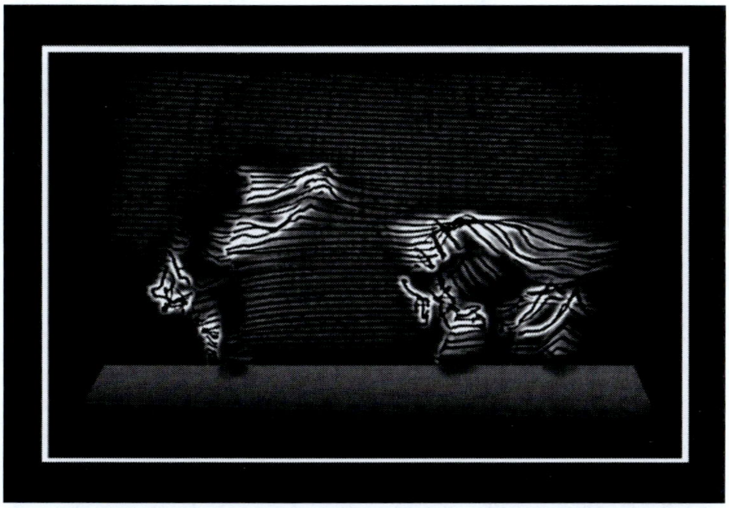

Figure 15. Example of using an infrared camera originally intended for gaming but now for movement and its realtime visuals for an audience (PrinceMio, 2012).

While there are numerous performances of kinect-MoCap-triggering visuals available on the the world database, the world-wide-web that illustrate this development, *Flow 1* is a particularly good example (PrinceMio, 2013). In this work, the individual seems to disrupt a series of lines that are dynamic, like smoke, by body position in space. The lines appear to adapt to the body mass. The technology appears to be easily accessible, with numerous tutorials freely available, including *Kinect: Projector Dance* (Figure 15).

The performance space seems limited to that of the approximately ten-foot wide screen, and the performers seen in these last two figures do not use much downstage movement. There could be various reasons for this. One might be aesthetic. Because the projection is coming from the front, the performer approaching the camera could create a dark silhouette on the screen where the projected light would be blocked. This would not be particularly objectionable were a viewer aligned with the projector and performer. But, with audiences numbering more than a single viewer, that is not always possible. Another reason might be technical. The Kinect was devised for the dimension of a room that one might be in when playing a game, not a large space such as a theatre. There is a defined range in the depth of field that the device can sense. To get around this limitation, one could use several Kinect devices to monitor motion, but they would also need to be calibrated together, as one does with traditional optical MoCap systems.

There are many types of visual and aural software connections, such as *Pure Data*, with which Kinect can be used in realtime. One specific example is *A Tale of June* (Figure 16). Here, using Kinect via OSC, an individual can control video in *Quartz Composer* and sound in *Max/MSP*. At one point, this example illustrates movement of an individual in a box. This shows an additional artistic functionality, one that includes objects moved by the performer and tracked in realtime. Also, this example makes use of a scrim on which the projection occurs from the front, and the performer behind it also affects the visual experience.

The Kinect can also be used to capture lateral movement combined with other devices such as a gyroscope and radio module (Xbee) to transmit speed of rotation (Figure 17). In this application, two sets of motion-related data are processed by software. The movement is minimalist (essentially swaying and spinning side to side) and the visuals are likewise simple (a wireframe sphere that moves side to side with the dancers and enlarges during certain spins). Realtime works using the Kinect sensor demonstrate many levels of experience and collaboration, and this example might be considered to exhibit an "entry level" of sophistication.

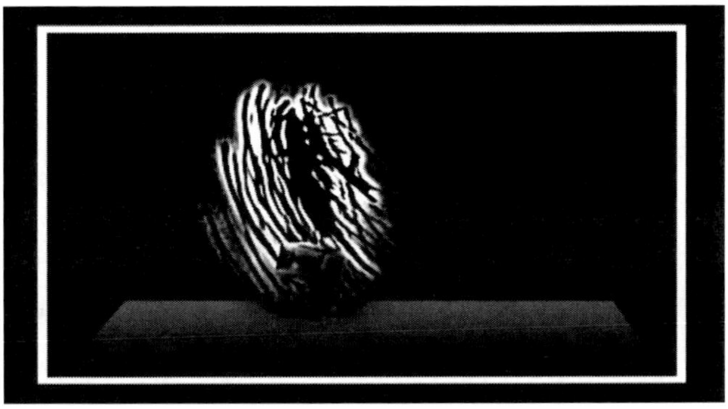

Figure 16. Example of another combination of software and hardware with an inexpensive camera to track movement, and the projection is on a scrim (Yoon, 2012).

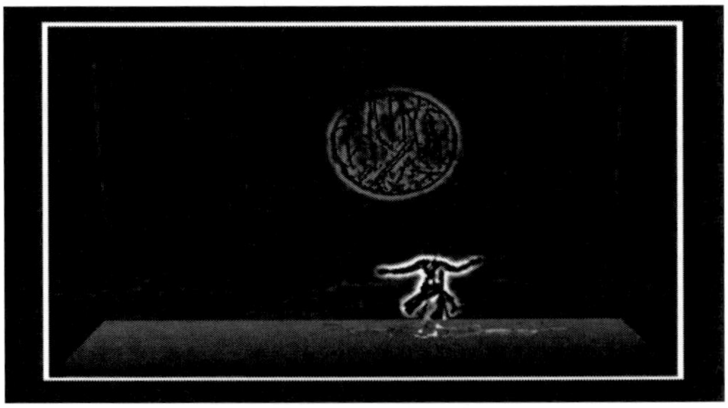

Figure 17. Example of multiple ways to get movement data to control visuals during a performance (Ghaderi, 2013).

The paper's opening teaser example of *Mr. & Mrs. Dream* uses several Kinects (Starr, 2013). As well, there are three projectors on the floor and three at the sides of the stage. Perhaps it is one of the most useful examples of a stream-of-conscious flow of visuals, as the two dancers on stage perform absurdist playwright Eugene Ionesco's surreal story about two characters who have escaped from the mind of their author and are journeying through his plays. High quality photos of some imagery and the set build and the setting of the projectors can be found at *Wired* (Stinson, 2013). It is noteworthy that, in the two cited reviews above, there is a sense of awe at the illusion, and of the virtual reality system.

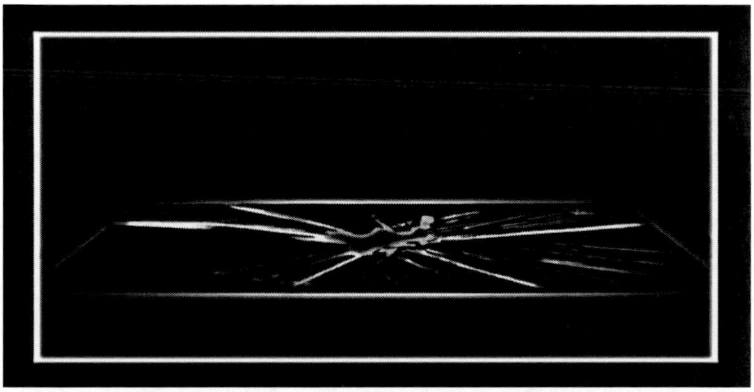

Figure 18. *Trinity*, example of another combination of software and hardware as well as top-down projection (Prosthetic, 2013).

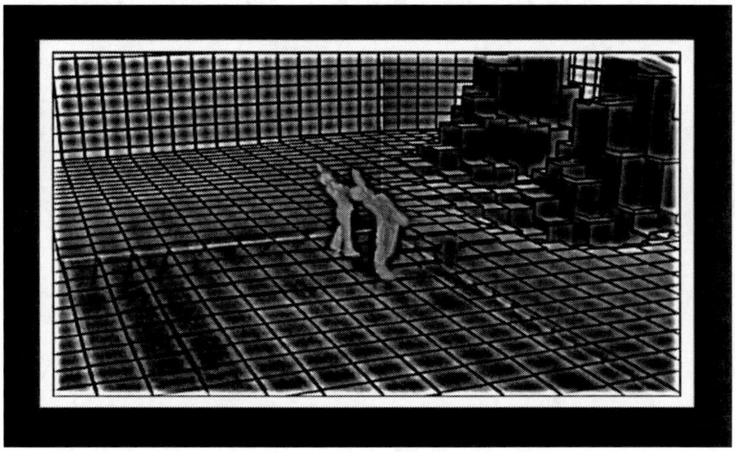

Figure 19. *Seventh Sense*, demonstrating projection mapping and illusion on four surfaces during a performance (Anarchy 2011).

In *Trinity* (2013), a videocam with infrared lighting is above the stage, and other than xy positional tracking, the mediated parameters are forces, directions, accelerations, velocity, and body area (Figure 18). The software here is *Max/MSP/Jitter* with the computer vision library CVJT by Jean-Marc Pelletier.

Visuals of *Trinity* are projected on the floor surface as noted in several other examples, but visuals projected on various surfaces seem to create more of a sense of immersion. Projection mapping as is in *Mr. and Mrs. Dream* allows a contiguous scene if desired on multiple surfaces. Such is also seen earlier with more abstracted visuals in the 2011 *Seventh Sense* (Figure 19). This kind of mapping allows realtime visuals to change and still maintain the illusion of depth and surround since it is seen projected on the floor, back and connecting sides of the stage. It has the overall appearance of a box with one side removed, through which the audience observes the action. The box appears to be about twenty feet on each side, and not only do the individuals in the space affect the realtime visuals, the audience members' movements also affect how the experience unfolds. In the visual example proved below, the performers are actually standing a few feet from the scrim on the upstage left corner and, if you look carefully, you can see a shadow on that scrim from one of the performers.

### Realtime Mapping of Motion to Visuals on a Non-Theatrical Stage

While most of the examples cited above might be considered high or serious art meant to engage an audience, sometimes capturing motion and creating visuals has a lighter, or possibly commercial purpose. A 'show' visualized in Figure 20 was part of the experience in the TOT Pavilion for the ITU Telecom World 2013.

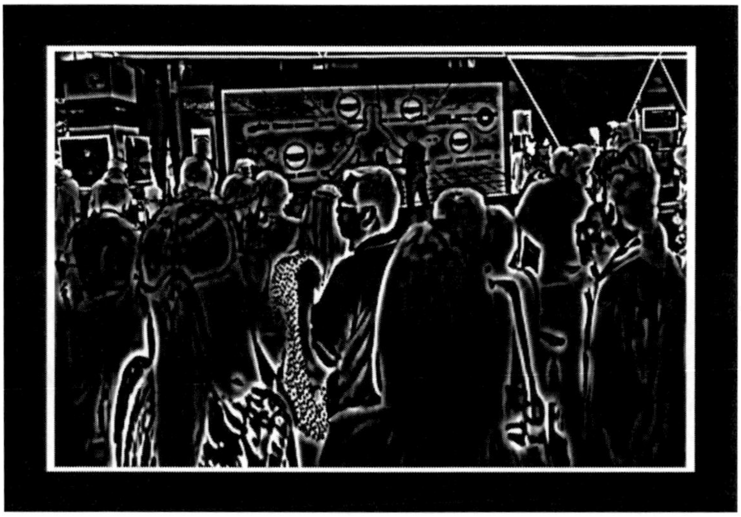

Figure 20. Example of an interactive performance not in the context of the theatre (Any.i, 2013).

## Realtime Partial Control of Aesthetic Experiences For Self

### Realtime Mapping of Motion to Visuals That Are Costumes/Clothes

Individuals can experience different clothing combinations without even trying them on if they position themselves in front of a motion tracking device and the output matches their body shape seen in a mirror-like screen (Figure 21).

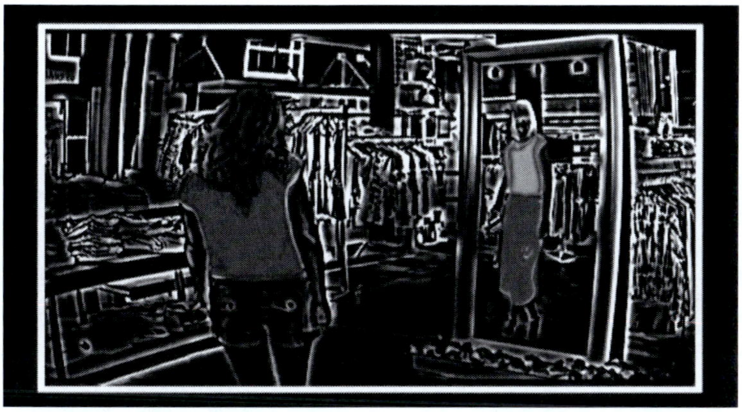

Figure 21. Movement sensing for application related to fashion (Kinect for Windows, 2013).

While this may seem purely commercial at first glance, it has some implications the author perceives. Fashion design is considered by many to reside in the realm of the artistic. There are sensibilities that determine choices dealing with proportions, shapes, colors, and other aspects of visual presentation and identity. While an aesthetic experience can be time-based, as in many of the examples above, it can also be for a moment. Maybe that moment where someone sees herself wearing a purple hat, an orange ascot, a black bodice, and brown pants might be an epiphany.

## Realtime Mapping of Motion to Online Virtual Characters/Avatars

Earlier in this study, MoCap standalones and MoCap libraries were mentioned in connection to an online virtual community, Second Life. It is also possible to stream MoCap data into the environment directly to a selected avatar (Figure 22).

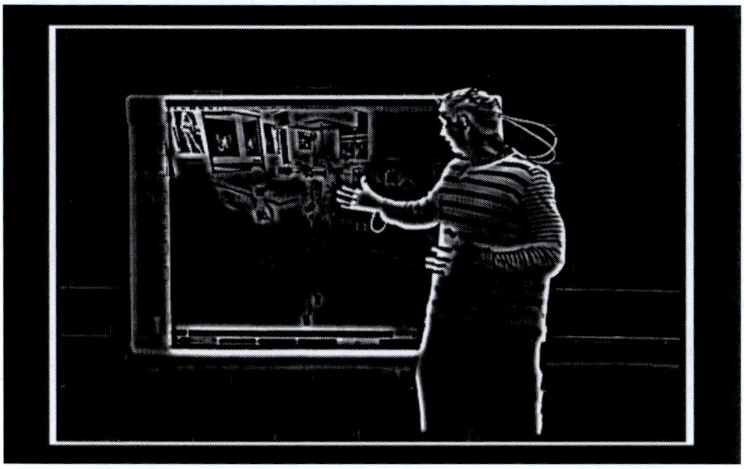

Figure 22. Example of realtime participation with movement in a virtual world (Ozzeuk, 2010).

This makes for the possibility of acting freely within an online virtual world using movements generated in the real world, and doing so without keyboards mice, or other traditional peripheral devices. And if applications don't already exist, this study predicts that there will soon come a time where avatars can interact in a realtime using audial communication rather than typing in words. And that this process will include a realtime means to translate from one language to another.

Is this biomechanical gesturing category one that could lead to an aesthetic experience and thus should it be included here? We could presume that one avatar might invite another avatar to go to a virtual art museum. Or one avatar might suggest to another to go to a play or concert or another virtual event but driven by real people in different geographical locations. A stronger response is that maybe one real person is a dancer who wants her avatar to perform an expressive sequence of movements for others. In the virtual world of Second Life, there can be crowds. There could be audiences; but they may be desire by all movement creators. Maybe a person just might like dance for herself as an avatar in the middle of a square in the virtual world, and maybe that would be an aesthetic fulfillment.

## Realtime Mapping of Motion to Create Artwork

*". . . just immersing myself in space is more like a dance . . ."* said Keane, who had been an animator at Disney and was now using a MoCap device. There are some motion-tracking devices that record your movements with mediated devices, such as a tilt brush—handheld hardware that works with a VR headset (Figure 23). While it may be something to engage an audience, for instance, with the skill an individual might demonstrate, it certainly would evoke an aesthetic experience for the self to create any form of artwork. This might be especially true for sculptors or painters initially, but might eventually be a channel for any artistic person to try things out.

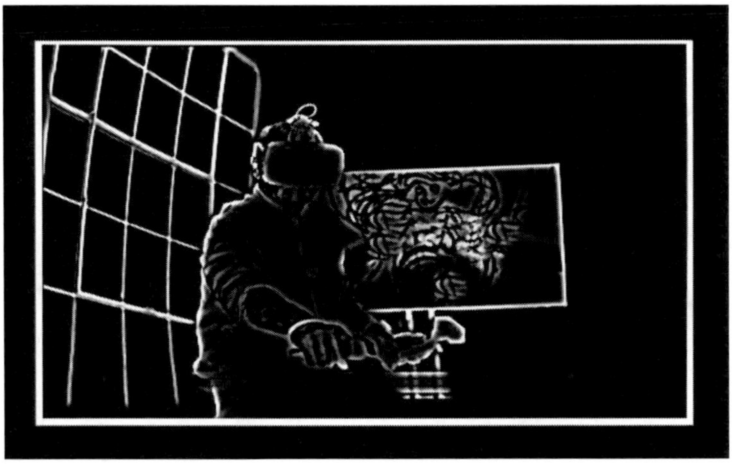

Figure 23. Example using motion-sensing to create artwork (Rundle, 2015).

## Realtime Mapping of Motion within Artwork for a Personal Aesthetic Experience

To fully discuss the premise of this paper—that through motion detection [input] one can cause things to happen [output] in realtime—in the domain of installation is a discussion beyond the scope of the author's intent. Rather the 2012 installation *Rain* might serve as an example (Figure 24). It is a rather energy-intense installation requiring pumps and a very specialized infrastructure of releasing 1000 litres of water per minute of the total 2500 litres through a gridwork of channels overhead. Essentially, the application of masking through analysis of a participant's position causes those channels above to turn off, as if the participant is walking under an invisible umbrella. Even though the installation is in a public area and an audience can watch a participant, the experience is truly for the individual who actually walks within the falling rain.

## Realtime Mapping of Motion to Create/Affect Visuals in Public Spaces

Anyone can claim, as in the earlier discussion about installations where the motion generator individual is in an art-focused space such as a gallery, that instead, in a non-art-focused public space—such as an airport, office building, or hospital— an individual creating a visual response through movement is an audience event, merely because it is observable by others. The author's contention is that the act of moving and creating something in a public space primarily brings satisfaction to the individual, recognizing at the same time that the nature of the visuals possible might have non-aesthetic, social, political, or commercial ends or intent. In other words, it is very unlikely that an individual says to herself, "I am going to interact with this motion-sensing system to entertain the strangers in this location." One example in this category takes place in a Bangkok airport where an individual can see herself moving within imagery of various locations in the world a person might like to visit (Figure 25). The interaction also includes being able to launch a lantern, and if one follows the link to media, one will see the pleasure the individuals receive.

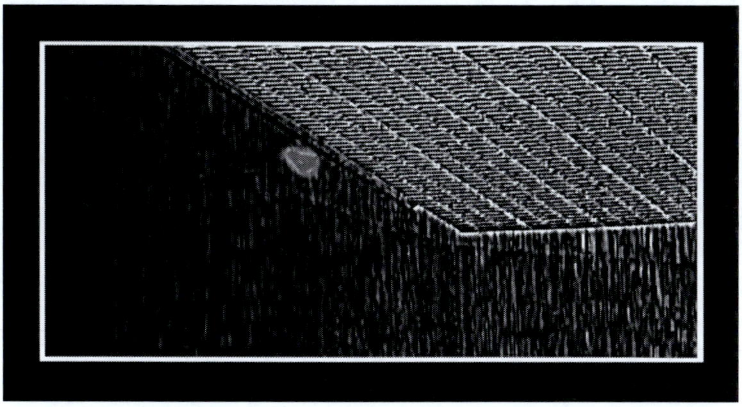

Figure 24. Example of a motion-sensing device that tracks location and turns off falling water so an individual stays dry while moving through the installation environment (Guardian, 2012).

Another example using twenty-four screens to create a wall is found in a Dubai business building (Figure 26). Screens have an advantage in that they can be in confined spaces, such as a passageway, as here, or in large spaces, as in the Bangkok airport, where there is no infrastructure to hang projectors or means of hiding them.

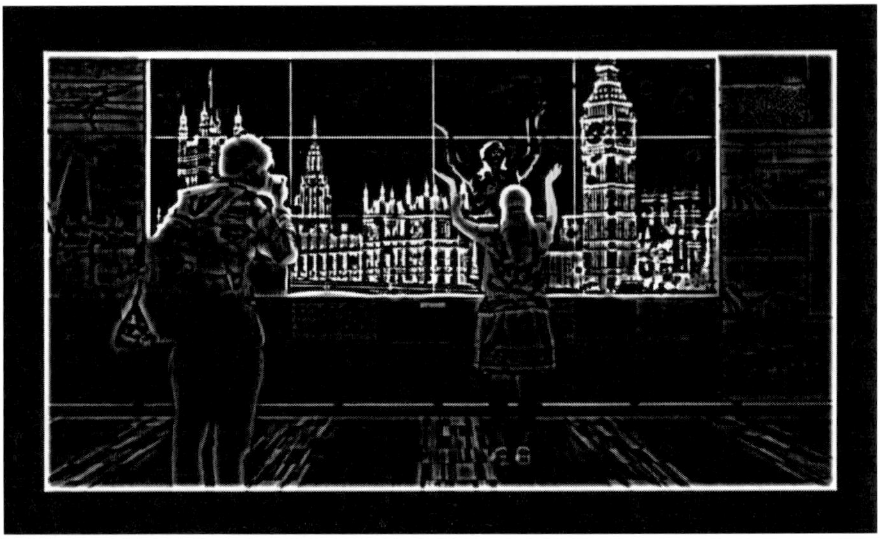

Figure 25. Motion-sensing that creates a mask with a video image to transport that image in realtime to a different location in virtual space (Apostrophy's, 2012).

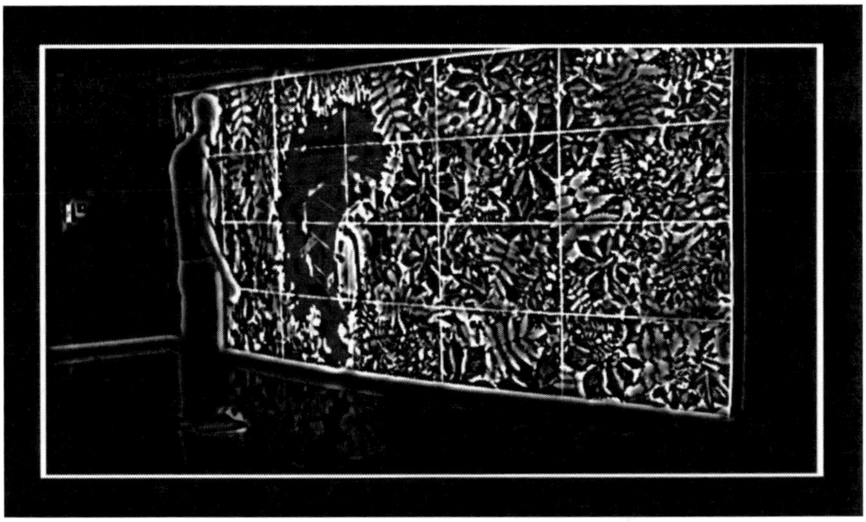

Figure 26. Example of a wall of screens that can allow persons in the space to cause changes in the imagery (But.nv, 2011).

A final example where motion affects visuals is located in a hospital, and contrary to the former two examples, does not have output on a long or wide series of panels (Figure 27). Rather, the output wall appears to be curved, and is quite high, with nearly the same height-to-width ratio.

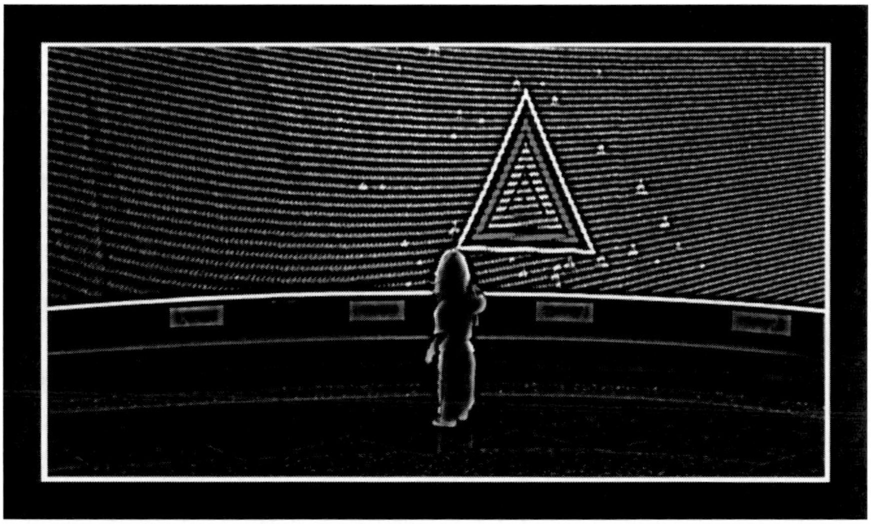

Figure 27. Example of a custom made surface in a public space for displaying imagery (Uconn, 2015).

## Realtime Mapping of Motion in a Room to Drive Visuals

Motion serves as the way to change the visuals in a room that has projections on three of its sides (Figure 28). The motion sensing device is positioned in the floor, so, in order to interact with visuals, one has to be at a specific location. The projected visual interface is not hidden; in fact, it is iconic. This example has three scenes, each one an actual movie: of an ocean with boats bobbing up and down, a landscape with turning wind power generators, and a city with fireworks exploding overhead.

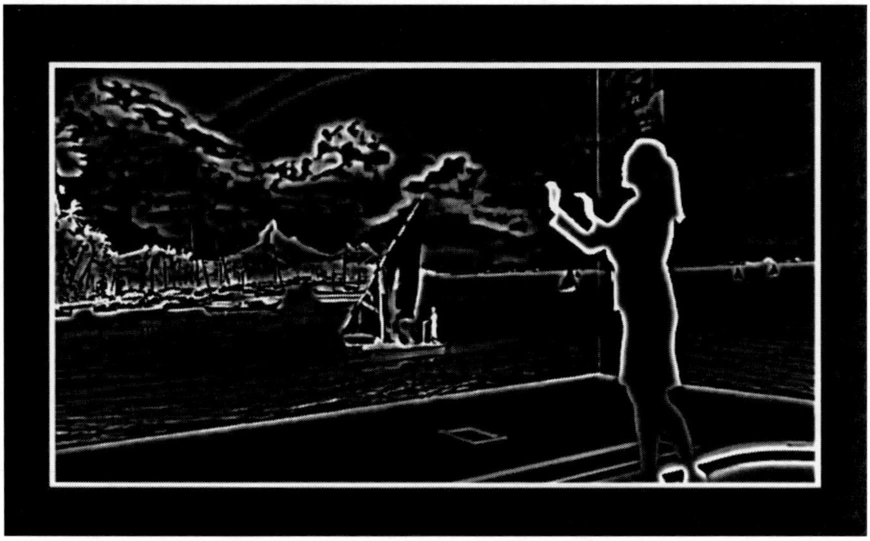

Figure 28. Example of a space where movement can control the imagery (Kenan, 2014).

## SPECULATION BASED ON OBSERVABLE CONTEMPORARY DATA

Through the sampling above of various applications, we have seen variations in the following: capture spaces, types of humans generating movement, out-putted sensory data, interface devices, and agency. The probable developments of each area can be proposed.

### Capture Space — For Audience and Self; Public and Private Spaces

For several examples above, the space for biomechanical gesturing had harsh 90-degree angles as the back wall/surface connected to the two sides. Since some places also had audiences, the fourth wall that would have created a closed system for projected visuals was left open for audience viewing. Caves—defined as areas of immersion where all six sides could have projection—predate this paper by a good twenty-five years (Evltube, 2007) though some consider the concept to go back much further ([N.A.], 2016). But twenty-five years ago, there weren't motion-sensing devices as we have today, or the advances technology has taken with projection and visual and sonic generation.

Public infrastructure related to this concept of a closed system has already existed for decades but with objectives different from those of realtime biomechanical gesturing. Who has never been in a planetarium watching moving imagery projected on a half dome? Who has never seen the Canadian Royal Mounties filmed in 360 circular [not spherical] degrees that was first presented at the '67 Expo (Highmore, 2016) or been in another venue featuring surround movies? These venues are solid, sturdy, made of concrete or other rigid building materials.

It is possible however that a closed semi-transparent fabric half-dome could be constructed, like a tent, where projection mapping could provide realtime visuals on the half sphere surface and on the floor based on movement-sensing within the dome. The audience could be seated outside of the dome, seeing [through a scrim-like fabric] the visuals and the performers on the inside. Or the audience could be seated inside around the perimeter. Theatres in the round go back further than Shakespeare's day more than 400 years ago.

Curved walls of screens already exist, and an enclosed facility with screens/monitors would eliminate some issues with projector throws. It appears fairly easy for DIY construction of flexible screens (ejunky, 2015) as many tutorials exist and are popular. That a screen floor surface could also be possible has already been realized in some motion-sensing applications (Mediatec, 2010). This could create a permanent structure that could serve a personal use or for dedicated use in a public space.

For many of any audience to a realtime event, the movement itself is certainly of interest, and each situation would have to articulate the philosophy of the relationship of the motion to the outputted visuals or sonic data. For instance, is importance of movement 80% and output 20%, movement 50% and output 50%, or movement 35% and output 65%? This might affect how both are treated and the spatial relationship of audience to performers.

Regardless, trends suggest an unwaivering march to illusionistic immersion in physical reality—space and time—using biomechanical gesturing.

## Performer — For Audiences, for Self

We have seen in examples above, where some who create biomechanical gesturing have had professional training to feel and control their muscles. This is especially true for those theatrical-oriented venues where people come to watch others create motion.

Yet there are some people that are very young, and some very old, who make events happen with their movement, especially in non-theatrical public places. As time progresses, similar to at first the trend of what demographic was first using a personal computer and then, what demographic was using the internet, there will likely be more people of all ages using their movement in immersive environments to gain self-experiences.

The author predicts that someone will likely develop a motion sensor input that tracks body and shape, and then outputs in a mirror-like projection, as in one of the above real applications already realized (Figure 21), but allows an individual to sculpt her body. So, for example, maybe the individual is a mid-life female who has gained too weight around her bum, or a mid-life professor with a belly; consider that each could push hands as if patting clay and make that torso look as it were when they were in grad school. The reflected image via the technology could certainly depict a differently proportioned body. Moreover, the programming of the experience could include that the person gesturing could touch a hotspot at a certain xyz location that had hair styles, and choose one to cover the gray or balding spots that have crept with the vicissitudes of life. This reflects 'vanity design,' and though the context borders on that related to social interaction, it, in this study, cannot be separated from a sense of aesthetics.

In one's own home, there could be panels in hallways that with sensing motion, creates various costumes on the torso, for instance, historical, such as Medieval, Renaissance, or Baroque, complete with simulated environmental images of castle walls with tapestries.

As well as inhabiting one's own body in the virtual reality, it is possible that one could inhabit one that were different from birth. Maybe a white person wondered how it might feel to move as one of a different color, or a young person might want to feel physically older and taller to match. Also in realtime, there could be motion sensing and associated visuals that could allow a female to see herself as a male, or vice versa. There are already algorithms that use the data of one's face and, through morphing processes, can reposition and shape aspects to meet the golden ratio, whether it be more masculine or more feminine. The technology to mask a realistic video image of a face already has been demonstrated in an example above (Figure 25), and with the speed of computing, realtime identity modification and placement in virtual settings that would have aesthetic design is reasonable, if not already obtained.

Further, a whole generation of primarily teenagers who have had motion-sensing devices as part of their gameplay consider motion and its effects on visuals more of a status-quo than a gimmick. While other motion-sensing devices, such as the Wii for gaming have been made available to countless millions of people, over 24 million Kinect units had been shipped by February 2013 (Kinect, 2016).

## Output — Now, Visuals and Sounds; Likely Others for Senses Other Than Sight and Hearing

Projection mapping is fairly mainstream with certain groups: indoor sports (see NHL, 2014) and night entertainment. As Ebner writes about stages, "The future is here." (Ebner, 2014). One of the problems for a researcher of a subject of realtime visuals is that one can search academic databases for published material but there is scant if any, and a methodology must be enlarged to seek primary sources online. Those involved with time-based work are rarely scholarly or academic-oriented. Projection mapping is also found in many public celebrations/events, even on public buildings. Crowds can be in tens of thousands. Certainly for an NHL hockey game, normal attendance is an average of 15 thousand. This means that more and more people are being exposed to the technology, and exposure brings a technology into commonplace acceptance.

Not only is the technology more affordable, projection mapping seems easy to accomplish, as there exist multiple user-friendly softwares and tutorials (see Resolume, 2013). Realtime sonic output driven by motion is fairly well established and requires less set-up and arguably, less expensive equipment, and will continue to trickle into mainstream. There may come a time when you saunter home and your house greets you with music based on your motion, and that music may be different every time you come to the door.

Many examples of audience-oriented experiences of sensory experiences (other than sight and hearing) can be found in the last few decades. One is R. L. Stine's *Haunted Lighthouse* that the author's daughters visited in the early 2000s (see BGWFans, 2013). While a somewhat traditional movie experience was projected on a screen, wind blew on the legs or back of neck, mist was sprayed, vibrations were felt, and it could have been perceptual, but salty air was smelled.

A scenario in the future exists that a person through her movements could cause certain scents that could be part of a dispensary library to be disbursed in the air. Imagine wanting to smell rosemary or crisp mountain pine at a certain moment in a sequence of biomechanical gesturing. And, if the scent isn't physically released, maybe a chemical that is identified with DNA/RNA or genes that evokes the scent is released. This and other predictions may seem outlandish, but who ever thought in 1950 that a human genome would be sequenced 51 years later? Imagine, for instance, the scents of a multicourse Italian meal, ending with chocolate, that might accompany motion.

With the progress of mapping genes, it might even be possible for a performer in decades wishing to incorporate a tactile sense of fuzziness to use movement to release a chemical to invoke that sense. In fact, there could be a dispensary library of airborne chemical-induced tactile experiences that might be part of an aesthetic experience.

Certainly the sense of hot wind or cold, or dry wind or moist could easily be triggered by motion-sensing output to appropriate appliances/fixtures/devices. In short, motions and their analyses could create richly layered and sophisticated aesthetic experiences. This could be for self and audiences alike. This could be in private or public spaces. This could happen in conjunction with LED screens or projections.

## Devices — Motion-Sensing Alone; or with Other Devices, Including Biosensors

This study has opted to primarily focus on immersion that does not isolate a person engaged in biomechanical gesturing from real space with the exception of the example of Keane drawing strokes in realtime. It was necessary to illustrate what could be accomplished by motion. There is a full rush currently to provide a 360-degree visual experience related to head motion. With all of the different VR headsets being developed for commercial release, this in itself argues that the desire for immersion seems to be a 'zeitgeit', or spirit of the times. People apparently want this kind of self-experience. People apparently want to be in their own space, their own room, and travel to, or be in other places, and interact with those virtual places on their own terms as much as possible. However an issue using merely a headset is that there can only be a limited physical sense of locomotion if any at all. Headsets come in two versions, tethered and wireless. While the visuals may create an illusion, if one is seated or standing in a room, not walking or running to move within the virtual world makes the experience seem [more?] counterfeit.

With a large amount of resources—for instance, an optical MoCap system, its operator, a dedicated large space, a grip to follow the person wearing a headset that has cables suspended from above, and more—*The Cell* allows natural locomotion [within bounds] (Manthorpe, 2015). We again want to emphasize here, again, that this does not create an audience aesthetic experience.

Currently there are other strategies, for instance one that lets a person with a VR headset slide their feet on a surface that senses that motion (Wainwright, 2014). In a sense, it is substitution of the real motion for one that requires a suspension of disbelief. A personal room's limitations to actual movement might be at best twenty feet, and similarly, in a large performance space with multiple motion-sensing devices, any straight-line or –path actions are an obstacle that predicted performance spaces won't solve without some adaption of a convention, or, possibly the introduction into the space of a floor-level treadmill device. Such a device would do three things, 1) reduce the amount of unrestricted floor space, 2) cause any straight-line locomotion to occur in a singular area within the sensed and rigged, outputted space, meaning that, 3) for instance, if a person wants to run for a mile North, and then walk for a mile West, the orientation of the visuals would have to change, as the treadmill would likely be fixed. This might solve a self-orientation issue, but not one for an audience. But bipedal locomotion is merely one of many actions one might wish to execute; what happens when one wants to fly, by one's own actions, not merely on dragons—as is visualized in the multiple award winning feature films *Avatar* (2009) and *How to Train Your Dragon* (2010)?

Two examples (VR reporter, 2016) and ([N.A.], 2015) illustrate this category. Both require the person creating movement to lie face down, but the latter reinforces this section of the paper's mantra with "To intensify the embodiment we include additional sonic, olfactory, tactile, wind and bio feedback." The human executing the motion must use arms and hands to create motion that maps to the visuals, as a first person POV. Reminds one of the 1963 Disney-animated *Sword in the Stone* where Merlin teaches Arthur, as a sparrow how to fly, doesn't it? Human desire for freedom has not been period-limited. [Over 50 years ago, they didn't even have personal computers; who might have imagined that a human might be able to experience flying, feeling the wind, with real motions, right in their own living room?]

This example generates another decision a producer of a future event may face, what is the POV? In the earlier example of the person wanting to run through a virtual reality, can the person shift from a first-person POV to seeing herself [3rd person] within the virtual environment, furthermore, how does one shift from near to far? The conventional iconic means of shifting perspectives near-far in several popular software applications is a magnifying-glass icon with a + sign to go closer and a magnifying-glass icon with a – sign to further away. But how is this done if a non-iconic interface is desired? How do we hide interface?

The last-cited motion-sensing example mentions biofeedback, and to accomplish everything related to realtime biomechanical gesturing and also discuss the sensing devices in this bio-sensing category is outside of what one can reasonably do here. But we can surmise, based on how research is going in multiple disciplines and predict how certain trends may converge.

Currently, there are investigations in the areas of imagination within the brain. Imagination is thought to be a function of the hippocampus (Zeldman, 2016). As well, there could be various sensors that are not invasive and could provide data from human movers that could shape the course of realtime outputted aesthetic assets (see Pinnacle, 2016). It is possible, with other devices, that through monitoring realtime components of human secretions such as sweat, or brain waves, that one could better understand the complex relationship of the imagination (see Pappas, 2014). It is possible that with analysis of certain biomechanical gesturing together with data from other sources of the human mover, that states of emotion or states of mind could be determined. The signals could be translated to numbers in certain categories that a computer, really, in essence, a big calculator, could process/understand. Through well-conceived software, these numbers or combinations could translate to visuals, sounds, or other outputted data.

Looking at history, humans have been going deeper and deeper into what a human is, how a human reacts to others, what a human needs to be able to live a long healthy life. It likely won't be in the author's lifetime when the inner-life of the mind can be shared with others in realtime, but is on the horizon.

Why won't the scenario be that a human sits [and not moves] and only with biosensing devices projects imagery or other data for an aesthetic experience? First, yes, this can be done now to some degree. To what extent is the experience satisfying? This paper opened with a brazen teaser leading to a sketchy overview of humanity's steps to exerting control. The author asserts that there can be no imagination without a host body, and for a host body to exist, it must consume things with energy and chemicals needed to maintain the body and expel the byproducts. As stupid as this may seem, one must use action to feed and expel. It is impossible for a human to exist on her own without action. If you are self-reliant and you don't move, you are dead. Moreover, there is also a physical sense of domination. We cannot wish, and have it happen, that one can dominate a territory, as some of the Renaissance European kings may have wished; one has to either do so in person or have a surrogate be there. While the Descartes' famous quote 'I think therefore I am' has oft been cited, the author would like to propose instead 'I move therefore I am.'

## Agency — That the Performer at First Has Greater Control of What Data to Output; Then Is Empowered to Create Her Own Assets through Improved Tools and Interface

In many of the examples presented above where environmental imagery was projected on floor, back, sides, or any combination, a producer/director/choreographer/designer shaped the experience with a sequencing of events/scenes. There was a sense of a beginning, middle, and end of the sequence to create an aesthetic experience for an audience before the ending applause. The person creating the movement had partial control to be able to affect the timing of visual or sonic moments within an event/scene, even some leeway when to initiate the next event/scene, but because a creative work needs to fit within the larger context of a period with an audience, and often there are multiple performances of the same piece, the creative work becomes somewhat fixed. For instance, a performance with movement and visuals of *Glow* is recognizable; another performance of it will look about the same. Aside from the likely fact that the performer did not design the visuals or sounds, one might assert that several samples presented above are based on a linear model (see A below).

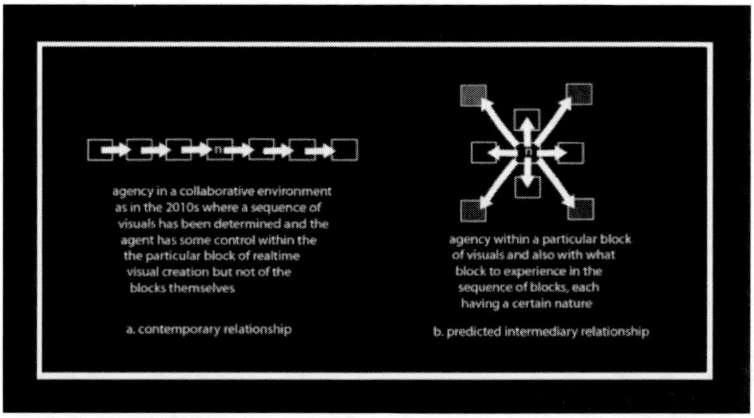


Figure 29. Schematic demonstrating the programming of aesthetic events and how the performer will be able to do more than merely trigger an event as a result of having branches of aesthetic sequences she can choose.

Starting in 2002, the author, also a professional educator, was challenging his new media students to create realtime performances with visual content—still images, animations, movies, words—by programming invisible hotspots on a projected screen and then triggering them, time after time, to branch into a unique aesthetic experience according to how the person performing the visuals felt about a certain piece of sound. In short, there were multiple pathways through the content and it might very well be that only a fraction of the prepared content was actually shared onscreen.

If for instance, the individual wanted more energy, she might click on an invisible hotspot in the upper right corner that would take her to certain sharp animations or bright colors or contrasting areas. Yet if at that moment, instead, she clicked in the lower right corner, then there might be a visual sequence that had more curves and pastel colors. In fact, there could be many hotspots/choices at any given moment. This is a paradigm of an interconnected branching system that could be modified to analyses of sensed motion (see B above). Rather than clicking on a hotspot as in the description above, an individual involved with biomechanical gesturing could go to a certain location in space to trigger a new branch of imagery. Or, there could be other ways through position/shape or other biomechanical aspects, such as acceleration to exert choice and control of the overall aesthetic experience. Using combinatorics, one can see above that if one has eight choices at the first event/scene, and eight at the next, and so on, that there could be thousands or millions of unique sequences of content according to the nature of the assets. The weakness of such a model is that the individual engaged in biomechanical gesturing has to remember what the 'triggers' are and furthermore, the correlated events to really assert control of her experience.

In our world where most areas require increasing specialization, the trend is also one of empowerment where once complicated procedures are simplified. For instance, one can use computer sequences that manipulate data by putting them as icons in the order desired without really understanding the syntax of computer languages such as C++. Or, for instance, with a click of a button one can change the look of a photo to simulate what an artist such as Van Gogh might have painted. If there is will on behalf of an individual to create her own asset libraries, certainly the means are available.

So returning to the opening of this paper, it is possible that an individual creates her own assets of the seasons, and has determined her own ways to use biomechanical gesturing to control those assets through motion sensing, and expresses herself in realtime with movement, imagery, and sound.

## REFERENCES

9GAG.TV (2016). *This is How Benedict Cumberbatch Became a Dragon in The Hobbit: The Desolation of Smaug.* Retrieved March 25, 2016 from http://9gag.com/tv/p/aAO6wM/benedict-cumberbatch-as-smaug-genious.

Anarchy (2011). *Seventh Sense* (2011). Retrieved January 5, 2016 from http://anarchydance theatre.org/en/project/seventh-sense/.

Any.i (2013). *Kinect Interactive Performance Show.* Retrieved February 4, 2016 from https://www.youtube.com/watch?v=UkjXYTjtkeo.

Apostrophy's (2012). *Kinect Interactive all Cotto: Share the World Pleasure @ Bangkok International Airport Donmuang.* Retrieved February 19, 2016 from https://www.youtube.com/watch?v=SHickY0NY-Y.

BGWfans (2013). *Official Haunted Lighthouse Video.* Retrieved March 26, 2016 from https://www.youtube.com/watch?v=LN1bAjCNS2s.

But nv (2011). *World's Largest Kinect Driven Interactive Video Wall.* Retrieved March 7, 2016 from https://www.youtube.com/watch?v=b2kqRv1WdEM.

Canal de Kinodanz (2010). Ghostcatching—Bill T. Jones. *Youtube.* Retrieved December 13, 2015 from https://www.youtube.com/watch?v=aL5w_b-F8ig.

Choensawat, Worawat, Minako Nakamura, & Kozaburo Hachimura (2013). GenLaban: A tool for generating Labanotation from motion capture data. *Multimedia Tools and Applications*, 74(23): 10823-10846.

Cultureband (2014). *Contemporary Dance: Chunky Move; Glow* (Official, Full, HD). Retrieved February 6, 2016 from https://www.youtube.com/watch?v=2AautwIOON8.

Dahl, Luke (2014). Triggering Sounds From Discrete Air Gestures: What Movement Feature has the Best Timing? *MIME 2014* (New Interfaces for Musical Expression).

Dancetechtv (2012). 16 [R]evolutions (2006) by Troika Ranch. *Youtube*. Retrieved 16 December 15, 2015 from https://www.youtube.com/watch?v=KFjy5Z2b-iE.

Dassault Systemes (2013). *Mr. & Mrs. Dream — "Magic Box" Technological Backstage.* Retrieved February 22, 2016 from https://www.youtube.com/watch?v=EB8JiFYRvwM

Delbridge, Matt (2014). *The Cooling Steam of the Polar Express: Historical Origins, Properties and Implications of Performance Capture*. PhD thesis, Creative Industries Faculty, Queensland University of Technology.

De Lahunta, Scott (2013). *Apparition; Klaus Obermaier and Ars Electronica Futurelab.* Retrieved March 11, 2016 from http://www.exile.at/apparition/background.html.

Ebner, Molly (2014). *13 Most Epic Projection Mapped Stages*. Retrieved March 25, 2016 from http://producerdj.com/2014/08/12-epic-projection-mapped-stages/.

ejunky66 (2015). *How to Build Your Own Flexible led Display XXL WS2811 Realtime Video Transmission 2000x1500mm*. Retrieved March 26, 2016 from https://www.youtube.com/watch?v=e-rdgB_19Fg.

ENRA (2013). *Enra "Primitive"*. Retrieved February 7, 2016 from https://www.youtube.com/ watch?v=IALr6M2NXsE.

Evltube (2007). *CAVE—A Virtual Reality Theater—1992*. Retrieved March 30, 2016 from https://www.youtube.com/watch?v=aKL0urEdtPU.

Ghaderi, Maziar (2013). *Dissolving Self: Wearable Technology + Contemporary Dance.* Retrieved February 25, 2016 from https://www.youtube.com/watch?v=llnSPaJdews.

GameSpot (2012). *Beyond: Two Souls — Motion Capture Video.* Retrieved February 18, 2016 from https://www.youtube.com/watch?v=_QZN2IC0vOo.

Griffiths, Jamie (2007). *Videmote.* Retrieved January 13, 2016 from https://www.youtube.com/watch?v=Ck5YYWiENjg.

Guardian (2012). *Barbican's Rain Room: It's Raining, But You Won't Get Wet.* Retrieved December 11, 2015 from https://www.youtube.com/watch?v=EkvazIZx-F0.

Highmore, Ben (2016). *Into the Labyrinth: Phantasmagoria at Expo 67*. Retrieved March 26, 2016 from https://books.google.ca/books?id=aal3Xm7WrewC&pg=PT147&lpg=PT147& dq=royal+canadian+mounties+360+film&source=bl&ots=3PLzs4XMKM&sig=tqG5wgTYsOvN675SnBBsi4J5tpY&hl=en&sa=X&ved=0ahUKEwimqvvYzuvLAhUN1GMKHRJzDdgQ6AEIJzAE#v=onepage&q=royal%20canadian%20mounties%20360%20film&f=false.

IDMIL-PI (2016). *Digital Instruments for Music and Dance Performance*. Retrieved January 29, 2016 from http://idmil.org/projects/prosthetic_instruments.

IDMIL-PM (2016). *Performer Movement Analysis and Synthesis*. Retrieved January 29, 2016 from http://idmil.org/performer_movement_analysis_and_synthesis.

*Interaction and Interfaces* (2016). Retrieved February 7, 2016 from http://artcom.de/en/research_focus/interaction-and-interfaces/.

Jurgensen, John (2011). Trying to Get *Tintin* Right; Spielberg and his Team Work to Avoid Some Pitfalls of Motion Capture in a New Adaptation. *Wall Street Journal* (Online) [New York, NY] 15 July 2011: n/a.

Kanal von 3DProjektion (2011). *3D-Projektion-Tanzperformance-Eventproduktion.* Retrieved 13 March 2016 from https://www.youtube.com/watch?v=pDBh4hF6p1A.

Kenan AK (2014). *Amazing Future Technology: Interactive Full Room Wall-to-wall Motion Sensor Entertainment.* Retrieved January 17, 2016 from https://www.youtube.com/watch?v=8fJ2ORNuwKI.

Kinect (2016). *Kinect from Wikipedia.* Retrieved March 8, 2016 from https://en.wikipedia.org/wiki/Kinect.

Kinect for Windows (2013). *Kinect for Windows Retail Clothing Scenario Video.* Retrieved March 1, 2016 from https://www.youtube.com/watch?v=Mr71jrkzWq8.

La Biennale di Venezia (2016). *Chunky Move* (Australia). Retrieved January 24, 2016 from http://www.labiennale.org/en/dance/archive/festival/program/chunky-move.html?back=true.

Macel, Emily (2010). *iDance.* Retrieved January 11, 2016 from http://dancemagazine.com/news/iDANCE/.

Manthorpe, Rowland (2015). *Visualise VR Have Invented a Full-body 'Cell" You'll Never Want to Leave.* Retrieved December 21, 2015 from http://www.wired.co.uk/news/archive/2015-10/19/visualise-vr-wired-2015.

Mark, David (2005). Realtime Motion-E Capture Makes Dance a Digital Art. *Electronic Design 53.10* (May 12, 2005): 19.

Meador, W. Scott, Timothy J. Rogers, Kevin O'Neal, Eric Kurt & Carol Cunningham (2004). Mixing Dance Realities: Collaborative Development of Live-motion Capture in a Performing Arts Environment. *Computers in Entertainment* 2(2): n/a.

Mediatec (2010). *Interactive LED Floor.* Retrieved March 28, 2016 from https://www.youtube.com/watch?v=g6N9Qid8Tqs.

N. A. (2012). *'The Hobbit'* and the Technology of Cinema. *Mint* [New Delhi] Dec. 18, 2012.

N. A. (2015). *Gamification in a Full Body Virtual Reality Context.* Retrieved March 20, 2016 from http://master.design.zhdk.ch/projekte/gamification-in-a-full-body-virtual-reality-context.

N. A. (2016). *Virtual Reality—Historical Overview.* Retrieved March 29, 2016 from http://science.jrank.org/pages/11561/Virtual-Reality-Historical-Overview.html.

NHL Video Center New Jersey (2014). *Devils Opening 3D Projection Show.* Retrieved March 25, 2016 from http://video.devils.nhl.com/videocenter/console?id=647349.

Ozzeuk (2010). Realtime Full Body Motion Streaming into Second Life. *Youtube.* Retrieved February 14, 2016 from https://www.youtube.com/watch?v=NFBv_ypyhiA.

Pappas, Stephanie (2014). *Will Wearable Tech Bring Humanity a 'Sixth Sense?'.* Retrieved March 21, 2016 from http://www.livescience.com/42490-wearable-biosensor-technology.html.

PrinceMio (2013). *Flow 1: Kinect Projector Dance.* Retrieved March 11, 2016 from https://www.youtube.com/watch?v=ISKV1BeB3pM.

PrinceMio (2012). *Projector Dance.* Retrieved March 7, 2016 from https://www.youtube.com/watch?v=FnulH8TrZVo.

Pinnacle (2016). *Biosensors.* Retrieved March 28, 2016 from http://www.pinnaclet.com/biosensors.html.

Prosthetic Knowledge (2013). *Prosthetic Knowledge Picks: Dance and Technology.* Retrieved January 2, 2016 from http://rhizome.org/editorial/2013/jul/31/prosthetic-knowledge-picks-dance-and-technology/.

Resolume (2013). *Awesome Set of Projection Mapping Tutorials.* Retrieved March 25, 2016 from http://resolume.com/blog/10750/awesome-set-of-projection-mapping-tutorials.

Rokeby, David (1986). *Works: Very Nervous System* (1986-1990). Retrieved February 23, 2016 from http://www.davidrokeby.com/vns.html [The author, Smith, experienced this system at a Dance and Technology Conference nearly 20 years ago. Rokeby's website lists a conference at Simon Fraser University in 1995, but I edited and published the proceedings of the conference actually held at York University in 1995.]

Rundle, Michael (2015). *Watch Disney's Glen Keane Drawing Sculptures in Virtual Reality.* Retrieved January 22, 2016 from http://www.wired.co.uk/news/archive/2015-09/14/glen-keane-disney-virtual-reality-drawing.

Sicart, Miguel (2015). How the 21st Century Will Be Defined By Games. *Wired Magazine.* February 4, 2015. Retrieved March 15, 2016 from http://www.wired.co.uk/magazine/archive/2015/04/ideas-bank/ludic-century.

Smith, A. William (2011). Mouvement Concrete; Modification of Biomechanical Motion Data for Non-narrative Aesthetic Time-based Expression. *Arts BioMechanics,* 1(1): 57-87.

StageNoise (2007). *Chunky Move: Glow*. Retrieved January 9, 2016 from https://www.youtube.com/watch?v=BVW92VR8n9M.

Starr, Michelle (2013). *3D Technology Brings the Stage to Life*. Retrieved February 24, 2016 from http://www.cnet.com/news/3d-technology-brings-the-stage-to-life/.

Stinson, Liz (2013). *A Virtual Stage That Bends Reality and Pushes Theatre's Boundaries.* Retrieved February 24, 2016 from http://www.wired.com/2013/07/this-virtual-reality-set-will-make-you-question-reality/.

StudioMyANIMATION (2011). *My ANIMATION Intro for Second Life*. Retrieved January 27, 2016 from https://www.youtube.com/watch?v=o0soUQlpCm4.

Talbott, Steven (2014). *MUST SEE!! Projection Ballet — Dancing Routine Lighting Effects (fx) using Projection DMX*. Retrieved February 8, 2016 from https://www.youtube.com/watch?v=JA6kl4RlA0I.

Troikaranch (2007). *16 [R]evolutions — About the Technology*. Retrieved January 21, 2016 from https://www.youtube.com/watch?v=Rbv7n0ZgA98.

UConn (2015). *Interactive Media Wall at Boston Children's Hospital*. Retrieved January 7, 2016 from https://www.youtube.com/watch?v=DD7gk2kHP3g.

Viennaexile (2010). *Apparition Klaus Obermaier & Ars Electronica Futurelab*. Retrieved February 19, 2016 from https://www.youtube.com/watch?v=-wVq41Bi2yE.

VistaAnimations (2010). *Vista Animations HUD-AO 3.0 — Introduction*. Retrieved December 28, 2015 from https://www.youtube.com/watch?v=GOtrPtF-Pe4.

VRreporter (2016). *ICAROS: A Flying Simulator That Trains Your Muscle*. Retrieved March 20, 2016 from https://virtualrealityreporter.com/icaros-virtual-reality-flying-simulator/.

Wainwright, Oliver (2014*). 'Full-body Joystick' Lets You Run and Jump in Virtual Reality Videogames*. Retrieved December 15, 2015 from http://www.theguardian.com/artanddesign/architecture-design-blog/2014/jul/28/full-body-joystick-virtual-reality-video-games-cyberith-virtualizer.

Weiss, Frieder (2010). *Mortal Engine by Chunky Move*. Accessed January 10, 2016 from https://www.youtube.com/watch?v=sbjOMualLVs.

Yanagisawa, Kazuma & Kakuro Amasaka (2013). Constructing A Scoring Support Approach Model For Classical Ballet: Combining Motion Capture And Statistical Science. *Journal of Business & Economics Research* (Online). 11(6): 241-250.

Yoon, Kiseon (2012). *Kinect Illusion — 2011/11/18,19*. Retrieved February 3, 2016 from https://www.youtube.com/watch?v=5Qjl4BQw8qw.

Zeldman, Peter and Eleanor A. Maguire (2016). *Anterior Hippocampus: The Anatomy of Perception, Imagination and Episodic Memory*. Retrieved March 16, 2016 from http://www.nature.com/nrn/journal/v17/n3/abs/nrn.2015.24.html.

Zimmerman, Eric & Heather Chaplin (2013). *Manifesto: The 21st Century Will Be Defined by Games*. Accessed March 15, 2016 from http://kotaku.com/manifesto-the-21st-century-will-be-defined-by-games-1275355204.

In: Advances in Arts Biomechanics
Editors: Gongbing Shan and Peter Visentin
ISBN: 978-1-53615-641-6


*Chapter 12*

# MOVEMENT SKETCHES, MOVEMENT SIGNATURES

***Gongbing Shan, PhD*, Tanya Harnett† and Peter Visentin#***
Faculty of Arts and Science,
University of Lethbridge, Lethbridge, Alberta, Canada

## ABSTRACT

For most people, the works of the Leonardo da Vinci easily come to mind as representative of intellectual inquiry contingent on an intersection of scientific and artistic modes of thinking. Both his scientific and artistic studies required analytic and creative skills that are indivisible. His work is combinatorial, not only exploring intersections of visual representation and scientific discourse but, more importantly, both qualitative and quantitative ways of thinking that connect them. Da Vinci's work describing the nature of the human form can be used to begin a discussion that draws attention to the overlapping interests of movement analysis science and the creation of visual artworks that comment upon the body corporeal. Since da Vinci, numerous artists have studied the figure by eye. His work provides a stage for explorations of space and perspective, and for their mathematical representation; something that may be thought of as "art & biomechanics".

A much more contemporary example of such an intersection can be linked to the advent of photography in the early 19$^{th}$-century, which revolutionized the way that people perceived and documented their interaction with the world. The camera changed everything. Photography provided a possibility of measuring space and time to be captured in imagery, which is a concept equally influential in art as it is in science. The timed multiple exposure and multi-frame photographic works of Eadweard James Muybridge define a clear boundary in this regard. The quantitative character of the medium and his careful documentation of Cartesian space made his images easy to be received by science. In those images where human (bio)mechanics are measured neither scientists nor artist could fail to be influenced by Muybridge's work. His work influenced new generations of artists as well as scientists. It inspired new ways to think about the embodiment of human movement in ways that included space (concepts of volume) and

* Corresponding Author Email: g.shan@uleth.ca; (403) 329-2683; Professor
† Corresponding Author Email:tanya.harnett@uleth.ca; (403) 394-3922; Assistant Professor
# Corresponding Author Email:visentin@uleth.ca; (403) 329-2782; Professor

temporal characteristics (the unfolding of an action). Fundamentally, this approach relies upon a viewer's experience, his/her familiarity with the condition of being a corporeal entity, to access subconsciously embedded empathy that is a necessary condition of human interaction within an environment.

As one more step in this direction, the current paper explores 3D motion capture technology as a tool for examining characteristic qualities of human movement. These qualities may be understood as defining human movement signatures. From a scientific standpoint motion capture technology can be used for generating data to analyze and understand these quantitatively. From an artistic one, discourse regarding the phenomenological aspects of these movements can be enhanced. In the following pages, multiple figures are presented as a means to demonstrate the utility of 3D motion capture as a tool to understand the construct of human movement signatures.

**Keywords**: 3D motion capture, full-body modeling, markers' tracing, drawing, gesture, line dynamic, sketching art of movement

## 1. Introduction

In today's world it is seemingly difficult to find an artist–scientist of the likes of Leonardo da Vinci (Abbott & Rutherford, 2005). This phenomenon is, in fact, a cultural gap; new technologies and disciplinary developments in the sciences and the arts sometimes can appear to foster apparently diverging interests and increasing emphasis on specialization. It is as easy for a scientist to forget that his/her understanding of science is as much influenced by the experiential as it is for an artist. Similarly it is easy for an artist to take for granted how his/her understanding has been mediated by technologies that have become a seamless part of 21st-cenury living. In Nature Supplement (Bohr, 1928), Niels Bohr laments that:

> "We find ourselves here on the very path taken by Einstein of adapting our mode of perception borrowed from the sensations to the gradually deepening knowledge of the laws of nature. The hindrance...originate(s) above all in the fact that...every word in the language refers to our ordinary perceptions."

Any increased awareness of this reality makes it clear that cross-disciplinary communication can be indispensable in the formulation of pioneering ideas about creativity in the arts and sciences.

According to Martin Kemp "to generalize about the relationship between art and science is not so much hazardous as impossible" (Kemp, 2005). However, as Bohr points out, artists and scientists share intuition based on experience. They also share metaphor as a means of communicating the particular as well as the universal. When a professor of physics attempts to explain cosmology by pretending that space is two-dimensional and that galaxies are formations on the surface of an expanding balloon, he/she is employing a metaphor that only works for those who have seen a balloon inflated, and it relies on the listener's intuition to extrapolate the imagery. When a painter creates imagery, the impact of his/her work on the viewer relies upon experience that can have relation to metaphorical, allegorical or linguistic constructs.

In terms of understanding the phenomena of and imagery associated with human movement, Niels Bohrs' comment remains relevant. Historical documents show both artists and scientists to have had a long interest in human movement (Shan, Visentin, & Harnett,

2010; Zhang, 2012). Clearly many of the foundational principles of artistic representation and scientific description in this regard can be traced to the early studies of Leonardo da Vinci (1452-1519) and the quantitative work of Giovanni Alfonso Borelli (1608 – 1679). With the advent of film technologies, the photographic work of Eadweard James Muybridge (1830-1904) marks a turning point in how visual representation of human movement could be understood. In the years since Muybridge, the sciences and the arts have experienced some perceived divergence due to increased specialization. Although new tools and technologies may be at the heart of some of this divergence, they also create possibilities of interdisciplinary interaction.

Artistic representation of various human movements often use gestures to suggest the essence of motion, while recent scientific explorations of human movement direct attention to characteristics, which can be considered quantifiable. As such, one might naturally inclined to think of these in terms of binary classification: the artistic representation of human movement is a creative act whereas from a scientific standpoint it is a calculated one. Clearly this overstates the reality but, in contemporary practice, many of the discernable identity conditions of being either an "artist" or a "scientist" reasonably lead the average person to sympathize with it. Even though the subject of human movement may be approached from motivational conditions that are disciplinary in nature, artists' and scientists' specialized understandings can be of mutual and beneficial influence. Both are involved in the construction/deconstruction of human movement patterns that may be understood as signatures of human movement. In this regard, technologies that have fostered disciplinary differentiation may also be used as tools of multidisciplinary integration. This paper aims to demonstrate one such example by considering the interstice formed between motion analysis science and art that is motivated conceptually and aesthetically. The links are provided by viewers' understanding of movement as something that is experiential as well as by the tapestry of language that has developed in order to communicate our existential perceptions.

## 2. Interaction between Motion Analysis Science and Creation of Visual Art

### 2.1. Leonardo Da Vinci – The Initiator of Biomechanics

Leonardo da Vinci easily come to mind as an intellect whose scientific inquiries were contingent upon his having creative and manual skills that might today be considered an artist's purview. With particular regard to da Vinci's works, the field of biomechanics owes a significant debt to his multidisciplinary talents - scientist/artist/inventor. Based on one late 20$^{th}$-century definition of biomechanics as a discipline - the study of the structure and function of biological systems using mechanical principles (Hatze, 1974) - da Vinci could rightfully be understood as the first biomechanist. His drawing *The Vitruvian Man* (Figure 1) is perhaps the earliest documentation elaborating the idea of human movement studies by means of "mechanics", even though it is more conceptually representative than it is scientifically consistent. The principles that the Vitruvian Man suggests are:

1) *Ideas of a reference system for motion analysis* – the drawing showed (through science and art) proportions that the body possesses in a circle's circumference pinpointed on the man's naval and the same man inside a square. This concept can be understood as the adapting of a perceptual and experiential understanding of a body into the schematic representation of that body. It does so using metaphor: the proportions and dimensions of circles and squares as the basic elements for understanding a coordinate system.

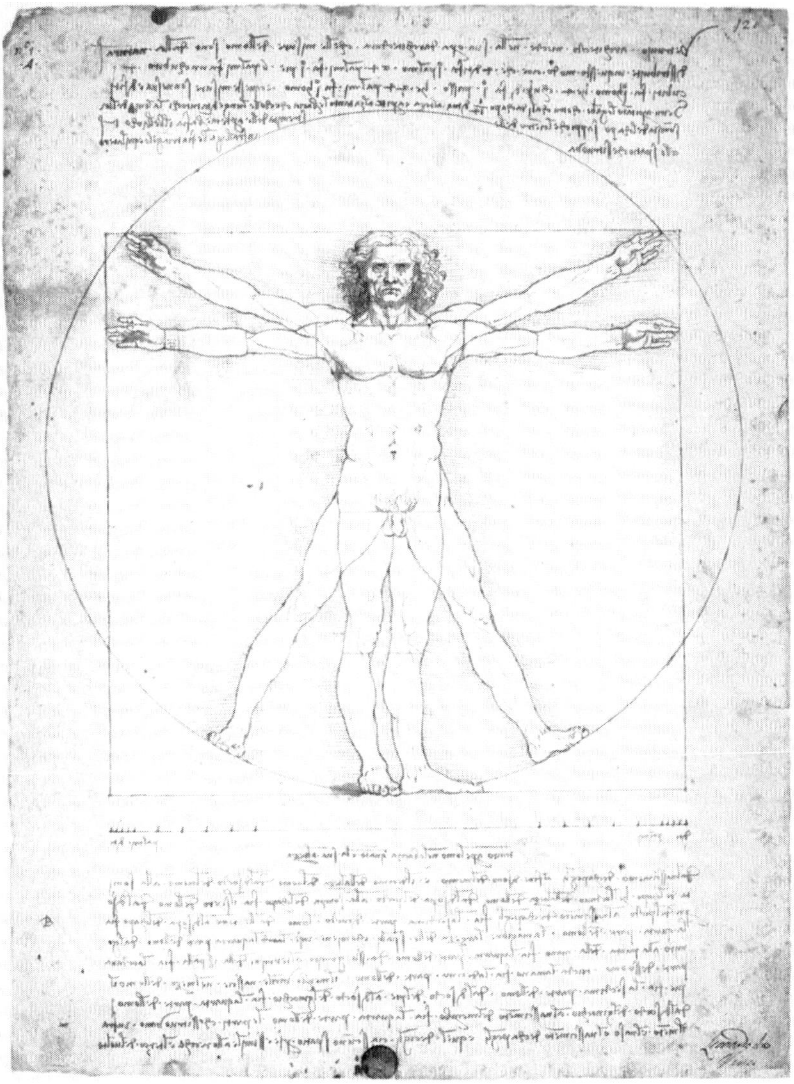

Figure 1. The Vitruvian Man by Leonardo Da Vinci (circa 1490).

2) *Conjunction of space and time* – Da Vinci was aware that, besides physical space, temporal dimension should also be considered in the studies of human movement (Veltman, 1992). He demonstrated this idea by the man's double pose, implying a man in motion. Artistically, Da Vinci's sequential-postural drawing technique gives us the feeling of movement and lays a foundation for today's movement analysis studies (Veltman, 1992).

In additional to his drawing of the Vitruvian Man, Da Vinci's other drawings and paintings illustrate his understanding of movement analysis. In his studies of flowing water (Figure 2a) the artist shows a portrait of a man, possibly himself, in contemplation of the turbulent motion that occurs as his walking stick is placed in the flow of the Arno river.

In both Figures 2a & 2b gestural aspects of the water in motion are depicted through the use of line, capturing a recognizable signature. On a two dimensional surface, da Vinci creates an illusion of a three-dimensional time-based phenomenon.

In other studies related to the human system, da Vinci applies his knowledge of anatomy to examine the human body as a mechanical system, which could not only be modeled but simulated (Veltman, 1992). Consequently, he made various drawings to demonstrate the system, e.g., principles and types of shoulder and arm movements (Figure 3). Simulation and modeling remains a fundamental tool in 21st-century biomechanics. Its models can simulate the body, its parts and its functions in ways that shed light on those elements of human movement and motor control that can actually be measured. Da Vinci aimed toward this concept almost 500 years ago. The interplay between science and art in da Vinci's works is undeniably meaningful. The characteristics of the period in which he lived encouraged: 1) scientific explorations as a guide for the creation of artworks (e.g., proportion & dimension as well as conjunction of space and time in painting) and 2) artworks that vividly demonstrate/visualize perceptual aspects scientific contents to viewers (e.g., showing human motor functions using anatomical/modeling details of a human body). In more recent times, the high valuation of specialization may be responsible for discouraging cross-disciplinary transfer of knowledge; polymaths such as da Vinci have become a rare commodity. As such, da Vinci's work should be acknowledged as a benchmark – perhaps the first interdisciplinary/cross-disciplinary approach toward human movement studies.

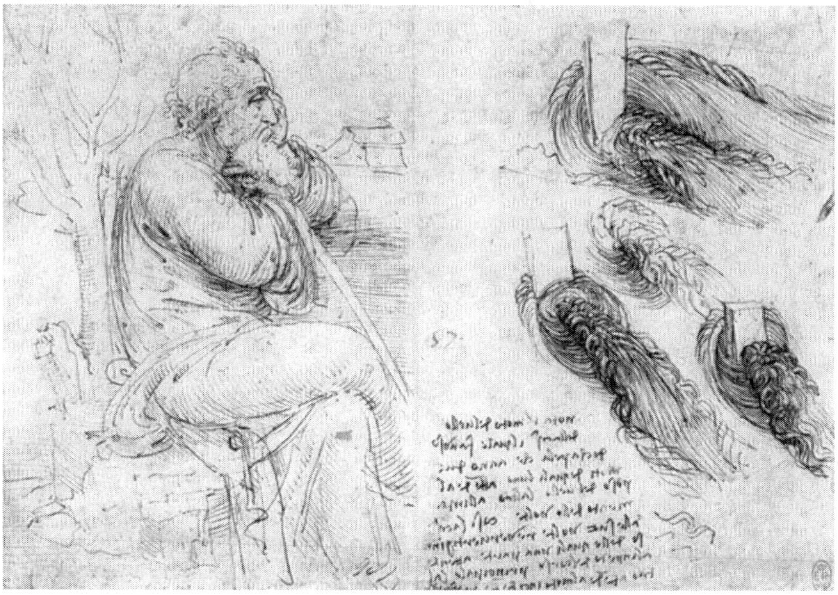

a)

Figure 2. (Continued).

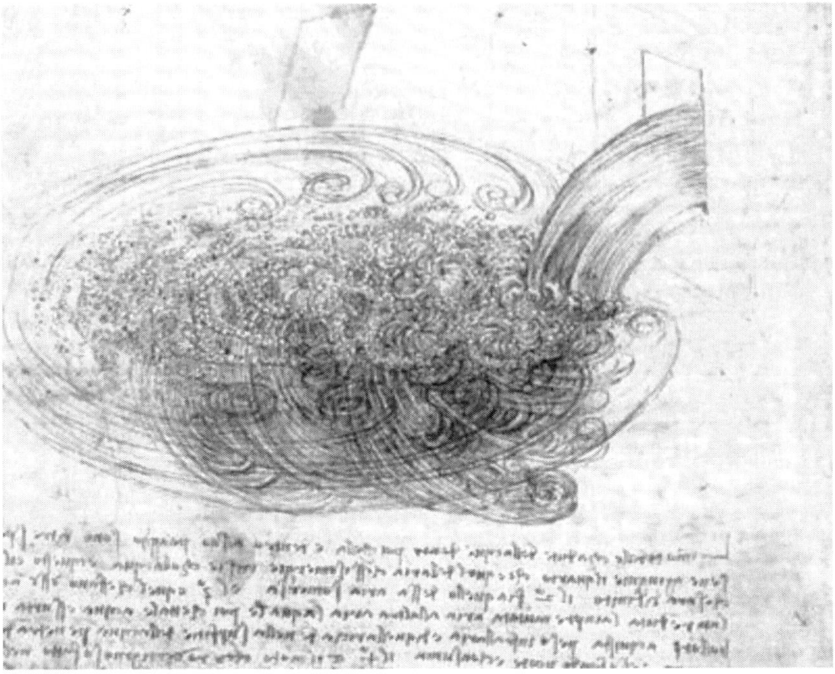

b)

Figure 2. Water studies by Leonardo da Vinci (ca. 1513).

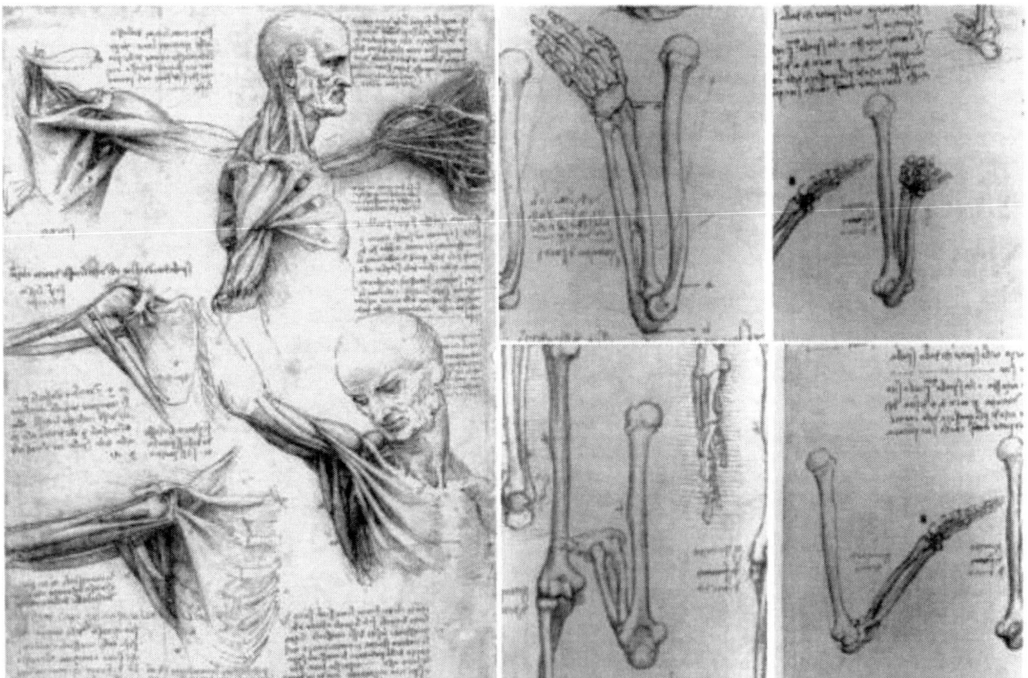

Figure 3. Principles and types of shoulders and arms movements demonstrated by Da Vinci's paintings.

Over hundred years later, Da Vinci's ideas of motion analysis were quantified into a Cartesian coordinate system by French mathematician and philosopher René Descartes (1596 - 1650) and into Newtonian mechanics (which studies the motion of systems in three-dimensional space) by the science giant Isaac Newton (1642 – 1727) (Arnold, 1989). Both of these quantitative methodologies provide the basis for classic mechanics, which is at the present time a foundational science in engineering and technology. Biomechanics, the study of the mechanics of biological motor systems, can be viewed as a sub-discipline of these.

## 2.2. Giovanni Alfonso Borelli – Pioneering the Mechanical Quantification of Human Movement

Giovanni Alfonso Borelli (1608 – 1679) is often described as the father of biomechanics (Pope, 2005). By applying principles of classical mechanics in his investigations of human musculoskeletal system, Borelli revolutionized the quantification of human movement. He dealt with the human body as a machine and showed that human motion could be explained in mechanical terms. As such, Borelli was the first to understand that the levers of the musculoskeletal system magnify motion rather than force, so muscles must produce much larger forces than any load resisting the motion. Notably, Borelli calculated the forces required for equilibrium in various joints of the human body well before Newton published the laws of motion (Martin, 1999). Lastly, Borelli quantified human movement, where our muscles generate forces that are transferred to the ground by bones and soft connective tissue. These concepts remain fundamental to modern biomechanics; in fact they are so much a part of the tapestry of our understanding that we rarely think about them.

None-the-less, the biomechanics of locomotion is remarkably complex and still incompletely understood (Ethier & Simmons, 2007). Borelli's research contributions in the field of biomechanics are contained in his two-volume book – On the Movement of Animals (De Motu Animalium), published posthumously in 1680 (Volume I) and 1681 (Volume II). The books are remarkable for their wonderfully detailed artworks (Figure 4 and Figure 5), without which biomechanical concepts such as locomotion, lifting, and joint equilibrium can hardly be explained (Ethier & Simmons, 2007).

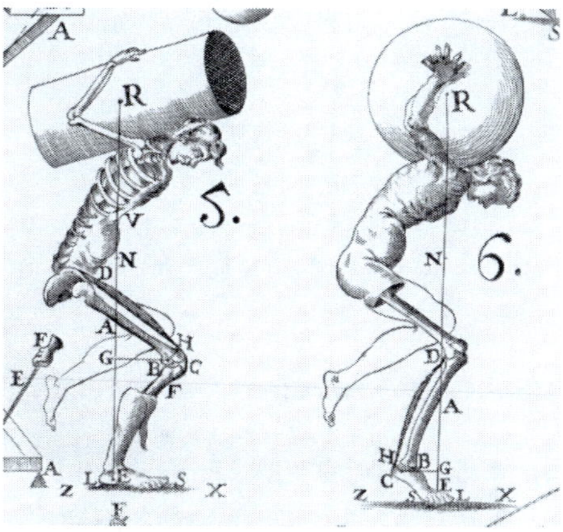


Figure 4. Two figures bearing loads. The illustrations demonstrate how the human leg is modeled as a twin-lever system controlled by muscles (from De Motu Animaliumby Alfonso Borelli (1608-1679)).

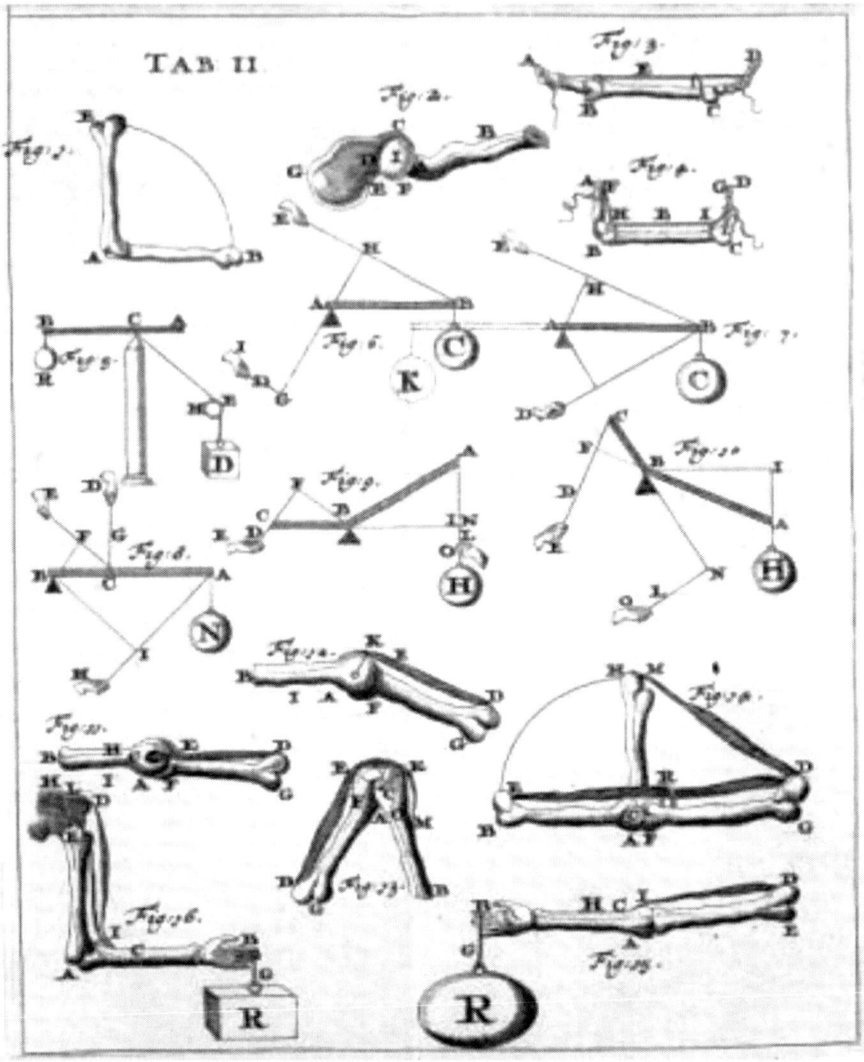


Figure 5. Borelli illustrated the biomechanics of the living body as a system of levers, pistons and pivots. This image models how the body suspends a load by a system of segments (levers) under tension from muscles (from De Motu Animalium by Alfonso Borelli (1608-1679)).

Clearly during Borelli's time, the integration of art and science, of metaphor and language, was expected. As a side comment, it is interesting to observe that disciplinary differentiation – a seemingly contradictory concept - arose in Borelli's century and the period marks the beginnings of conceptual divergence for the sciences and the arts (Martin, 1999).

## 2.3. Pre-Muybridge – The Art of Movement Representation

After Borelli, there is little sign of interaction between movement science and visual art in the science literature until Eadweard James Muybridge (1830-1904). Muybridge made several breakthroughs in his efforts to design new technology to undertake motion analysis. In visual art history prior to Muybridge representations of human movement appear to be stopped in time, suggesting a kind of mimesis that might poetically record or depict gesture in a composed 2D picture plane. Essentially, these capture a visual "affect" or a critical point during the unfolding of a gesture. When used as means to suggest the essence of motion, affects need not be accurate anthropokinetically. They may be merely representational, relying on metaphor, iconography, cultural loading and the experiential knowledge of the viewer (Shan et al., 2010).

Informed by science, da Vinci's drawings related to human movement became more realistic than earlier artworks. Although the idea of temporal dimensions was incorporated into his paintings (Figure 1), da Vinci still focused on visual aspects of realism in his images rather than a holistic representation of time-based events. Therefore, da Vinci's artworks that include temporal features can be categorized artwork inspired by human movement but not ones in a Cartesian framework. Borelli's main contribution was the instigation of the quantification of human movements. He used artworks in his study for one purpose only – vivid dissertation of scientific knowledge. Like da Vinci, his drawings were scientifically accurate and focused on details of human motor functions. Hence, the drawings in Borelli's works may be artwork that quantifies elements of human movement but they are not ones inspired by it. To quantify the visual qualities of human movement as in such a way that they become universally understood as a metaphor for human experience and not merely a descriptor of it, another approach is needed – one possibility is to uncover and portray movement signatures of that experience (Shan et al., 2010).

## 2.4. Eadweard James Muybridge – The Beginnings of Understanding Movement Signatures

If da Vinci's drawings suggested that movement could be illustrated by using conjunction of space and time, this idea was experimentally furthered by Eadweard James Muybridge (1830 – 1904) in his studies of locomotion. His photographic images represent profound changes in documenting dynamic, time-based events on a 2D surface. The camera changed everything. Many of his photographic techniques were of his own device. Muybridge was the first to systematically develop equipment and techniques to evoke the unfolding of human movement (Solnit, 2003). Multiple time exposure and multi-frame stop action imaging techniques differentiated his work within the discipline of photography. Thus, scientific and technological method, developed by a photographer, assert seminal influence on creation within the visual arts as well as revolutionizing movement analysis studies.

> Pre-Muybridge, affective representations of human movement using stop-action are clearly the norm. In 19$^{th}$-century Romantic imagery, movement might be alluded to in various ways using iconographic reference, formal structure, brush-stroke quality, painterly techniques, use of color, etc., or other culturally/stylistically loaded techniques. In images alluding to movement, "representation" involves synthetic, affective interpretation by artists (Shan et al., 2010).

Artists of the late 19$^{th}$ and early 20$^{th}$ centuries were highly influenced by photography. Muybridge's photographs using multi-camera stop action opened a new path for artistic creation. Edgar Degas (1834-1917) openly expressed his admiration for Muybridge's work. He studied and copied Muybridge's images before he "proclaimed his artistic ambition as the capturing of 'movement in its exact truth'" (Gonpertz, 2012). Degas was a classically trained draftsman highly influenced by Jean Auguste Dominique Ingres (1780-1867). Photography was a tool that he married to his classical understanding of artistic creation. Degas' images of dancers (1895-1896) are notable in this regard. He employs photographic source material and collages several images together to form a composition (Figure 6). This photographic montage informs the classical artist who not only renders the image but interprets movement within the drawing through the use of line.

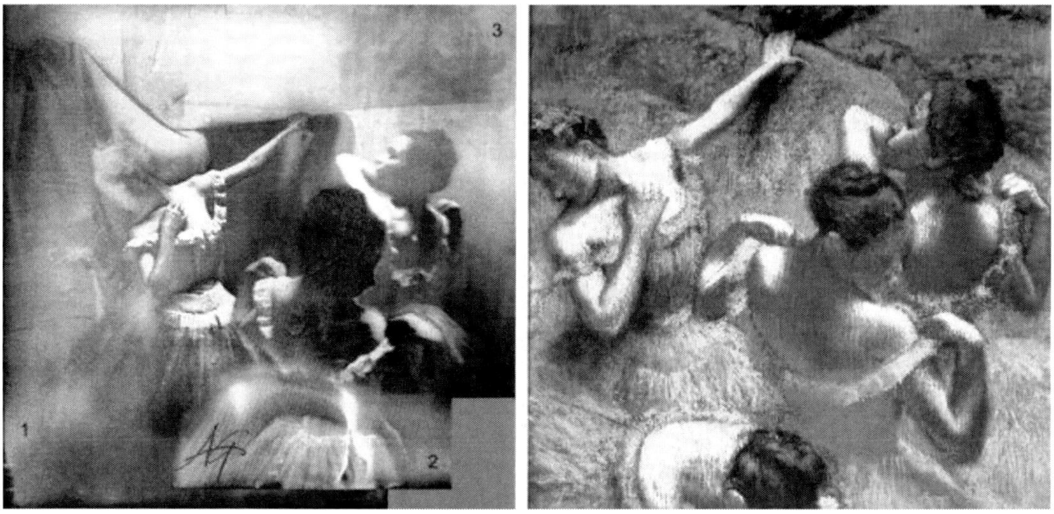


Figure 6. *Dancers* by Edgar Degas (1895-96).

Another representative example of science, technology and artistic interaction post-Muybridge can be found in Marcel Duchamp's *Nude Descending a Staircase* (1912) (Figure 7). One can clearly find elements of time exposure and multi-frame stop action similar to those of Muybridge's earlier image *Woman descending the stair* (Figure 8) and *Pole vault* (Figure 9) (Hunter, Jacobus, & Wheeler, 2004; Zhang, 2012).

It is impossible to overstate the significance of Muybridge's work. By relying upon the viewer's ability to integrate a time-based phenomenon into a perceptual whole, it fostered an ethic aimed toward a holistic embodiment of the human experience. His movement studies provide the essence of movement signatures - metaphors of the human experience that account for space (concepts of volume) and temporal characteristics (the unfolding of an action) as well as the intentionality of the act. All of these are embedded in single-frame, time-independent images (Shan et al., 2010).

The signature of bicycling – a drawing in Hay's (1985) biomechanics textbook (Figure 10) shows the uniqueness of that particular motor skill. It is a wonderful example, reminiscent of a Muybridge image that uses multi-exposure photographical techniques (Figure 11). The original photograph clearly identifies that the signature of bicycling should include two characteristics of the motor skill: 1) the translation of the movement, represented by the trunk motion, and 2) the rotatory movement, represented by legs and wheels. Hay's drawing is a cartoon of a movement signature.

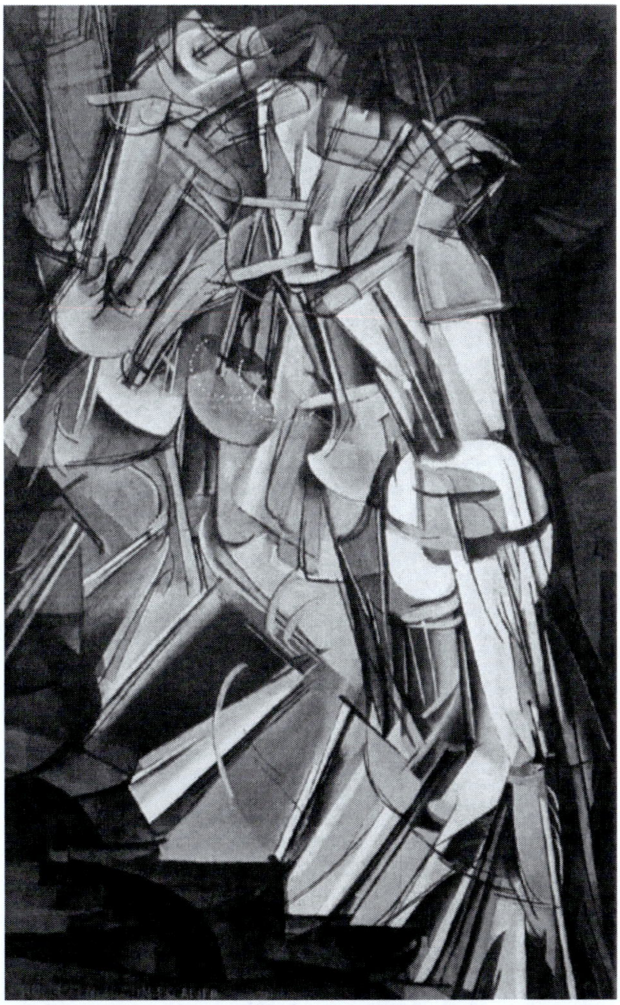

Figure 7. Nude Descending a Staircase (Marcel Duchamp, 1912).

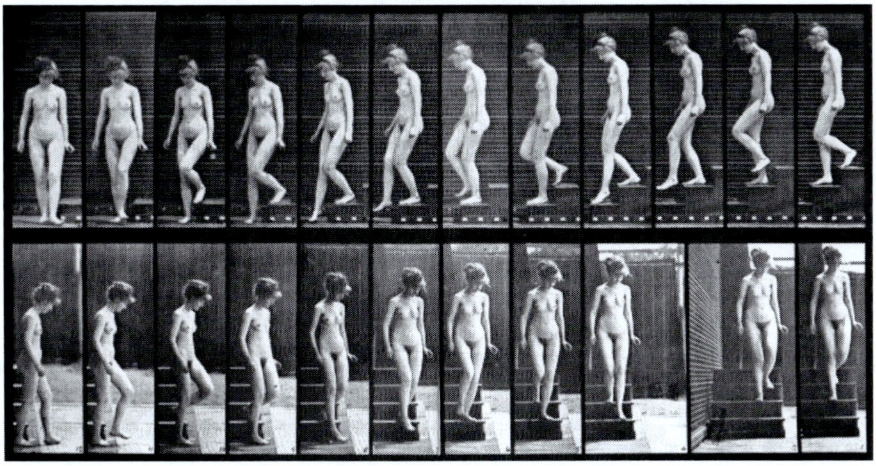

Figure 8. Muybridge's time exposure and multi-frame stop action image – Woman descending the stair (1880's).

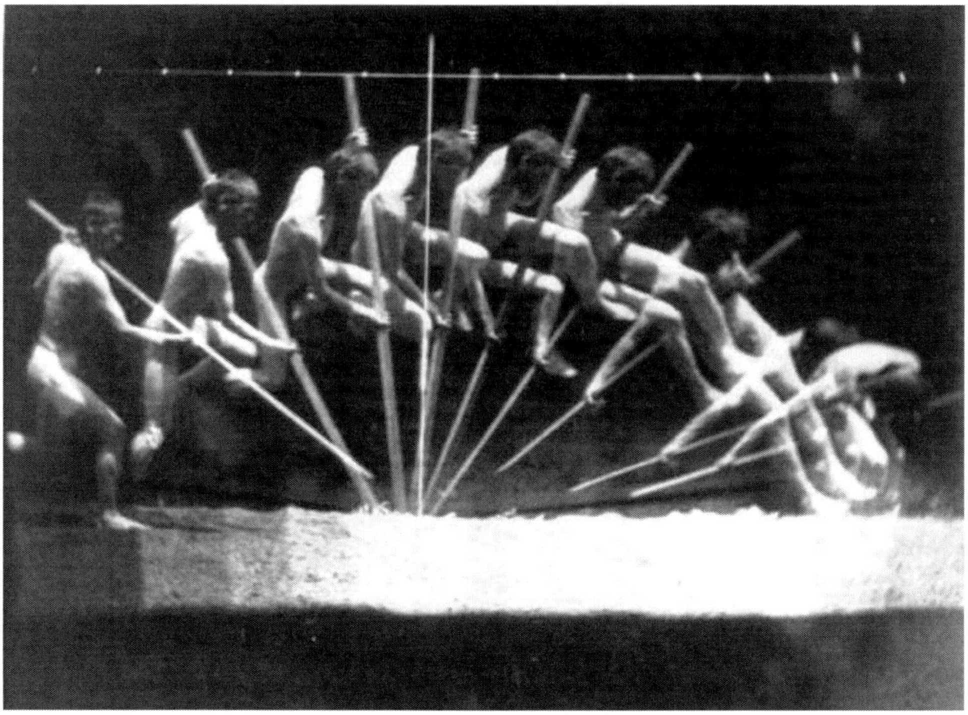

Figure 9. Muybridge's multiple exposure image – Pole vault (1880's).

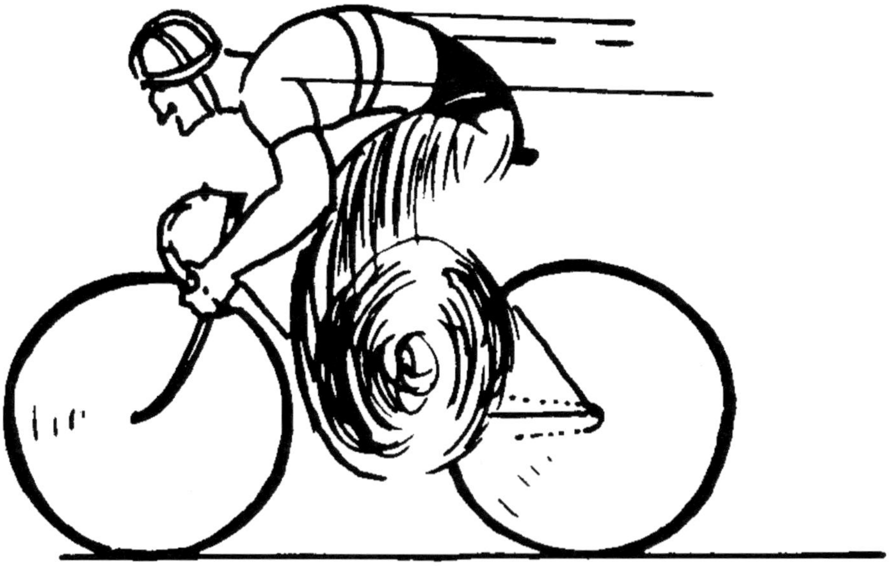

Figure 10. Signature of bicycling – translation [trunk] + rotation [legs and wheels] (Hay, 1985).

In fact, for some human movement/motor skills, multi-exposure images could be sufficient in and of themselves to communicate movement signatures. A good example can be seen in the image of a golf swing created by Edgerton in 1938 (Figure 12). It is well-known that golf swings are complicated movements that require coordination of all major body segments. The whip-like sequential joint controls of the trunk-arm-club system for maximizing club-head speed just before impact with the ball can be used to evaluate the efficiency of a swing (Shan, Zhang, Li, Hao, & Witte, 2011).

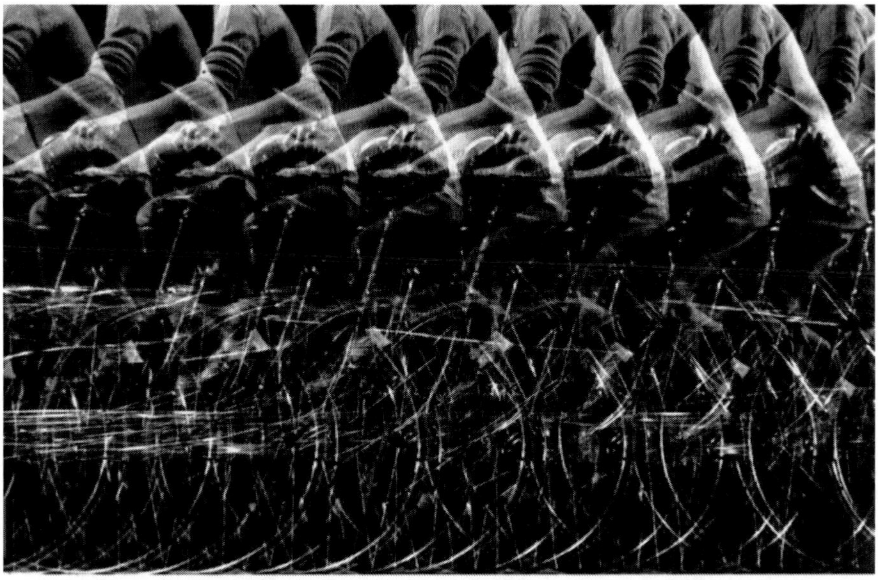

Figure 11. Image of bicycling made by using Muybridge's multi-exposure photographical technology (early 20th century, author unknown).

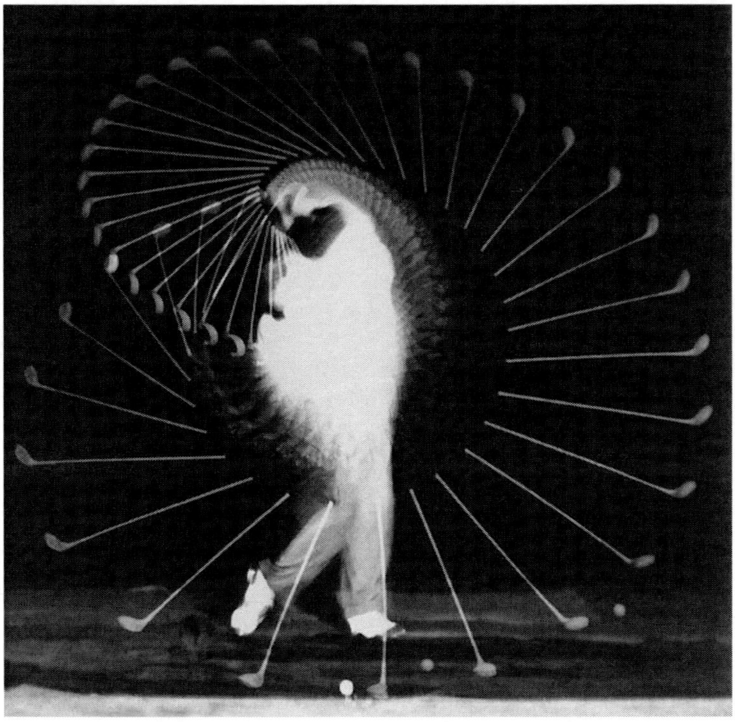

Figure 12. The signature of golf swing created by Harold Edgerton in 1938 multi-exposure photographical technology (Siegel, 2012).

Signature characteristics – rotations of trunk and arms for accelerating club swing as well as increasing its angular speed (shown by the angular acceleration, or the distances between club positions on the photograph) during the down swing for maximizing club head speed before ball impact – can be notably identified in Figure 12.

The above examples emphasize the reality that depictions of human movement have had the particular challenge of representing time-based events in a single image. Muybridge may have realized the idea by using multi-exposure photographic technology; however, his work was limited by the 2-dimensional measurement tools (photography) at his disposal. Some movement signatures can hardly be represented on a 2D surface. As motion capture technology is further developed, the three-dimensional aspects of human movement can be revealed and used to abstract perceptually characteristic elements - movement signatures.

## 3. Three-Dimensional Motion Capture as a Tool to Aid Description and Analysis of Human Movement Signatures

If photography tested the limits of visual culture in Muybridge's day, 3D motion capture does so in the 21$^{st}$-century. In Muybridge's work and the work of others whose images are created on a 2D plane, cognition of the dynamic world relies upon viewers' perceptions to compress information from multiple images into a summative experience, or to comprehend their metaphorical relationships. As an extension of film techniques, 3D motion capture permits detailed recordings of human movement in 3D computer space, resulting in data that may be used to create time-based linear trajectories. These can be used to determine patterns that are linked to gestural phenomena, notwithstanding the activity. In combination, these patterns become signatures, graphics that convey the essence of human movement and trigger a viewer's perceptual awareness. Such signatures allow us to effortlessly recognize complex human movement forms. They offer, via a short temporal snapshot, the identity of the movement without requiring conscious cerebral processing of its individual defining characteristics (Shan et al., 2010).

### 3.1. Three Dimensional Motion Capture

A 3D motion-capture system can be used to record full or partial-body movement. Multiple reflective markers need to be placed on a subject in a manner that identifies body segments and joints, points Borelli rightly identifies in De Motu Animalium as anatomically significant for human movement. Using motion capture, all movements of markers are recorded by multiple cameras. Data from these cameras can be cross-referenced to give marker position in a numerical coordinate system, Cartesian XYZ space. These can then be transformed into trajectories using computer imagery. Once captured, other information can be added to trajectories, such as surface meshes (resulting in animations of human movement) or computer-generated muscular-skeletal models (representing anatomical elements of human movements) (Figure 13). In terms of gestural communication and its representation, 3D motion capture technology supplies a quantitative and qualitative view of temporal events, something not possible with 2D technologies alone. Perceptions of gesture as an unfolding of time-based events are intrinsically integrated with the form and function of human movement.

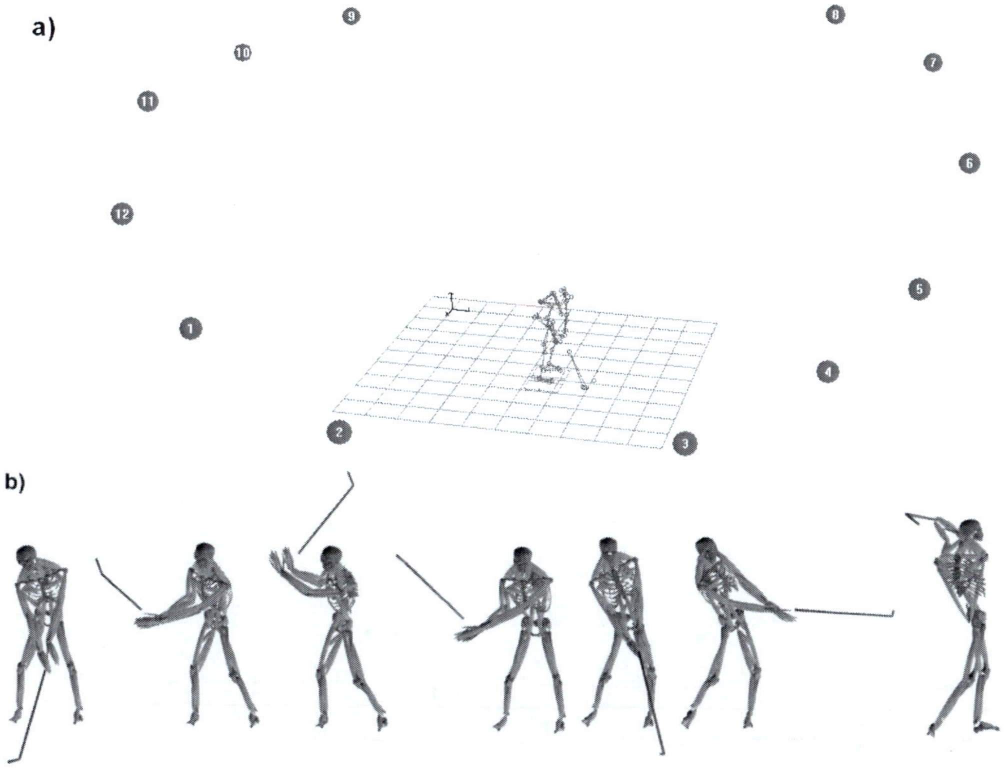


Figure 13. a. 3-D motion capture reconstruction showing multiple camera placement and a wire frame mesh reproduction of golf swing. b. Sequential stop-frame reconstruction of the movement.

Since 2D images that may be created from 3D information rely upon the experience of the viewer, and all experience is time-based, movement signatures must recall the temporal and experiential nature of the phenomenon. These are keys to the creation of movement signatures. Figure 14 shows a golf swing signature generated by 3D motion capture technology. Comparing this movement signature to the 2D movement signature in Figure 12, one can clearly identify that the fast rotations of trunk and arms for accelerating club swing and the speedy club-head movement.

It is interesting to compare the golf swing to a baseball swing. Biomechanically speaking, both skills employ similar fundamental mechanics: fast rotation. The only major difference is the trunk orientation. This difference is clearly visible in the movement signatures (Figure 14 and 15).

The rest of this paper demonstrates the use of 3D motion capture as a means to extract dynamic trajectory content and the temporal nature of selected human movements. This technique differs from photography in two fundamental ways: 1) the information gathered is not an actual image, rather it is a collection of computer-generated trajectories tracing pre-defined points, and 2) the viewer's choice of perspective is not controlled by camera placement, since it can be altered after the motion has been captured (Shan et al., 2010). Images synthesized from motion capture trajectories can be used to generate signatures of human movement – ones that are objectively representational of temporal and corporeal perceptions of bodily experience.

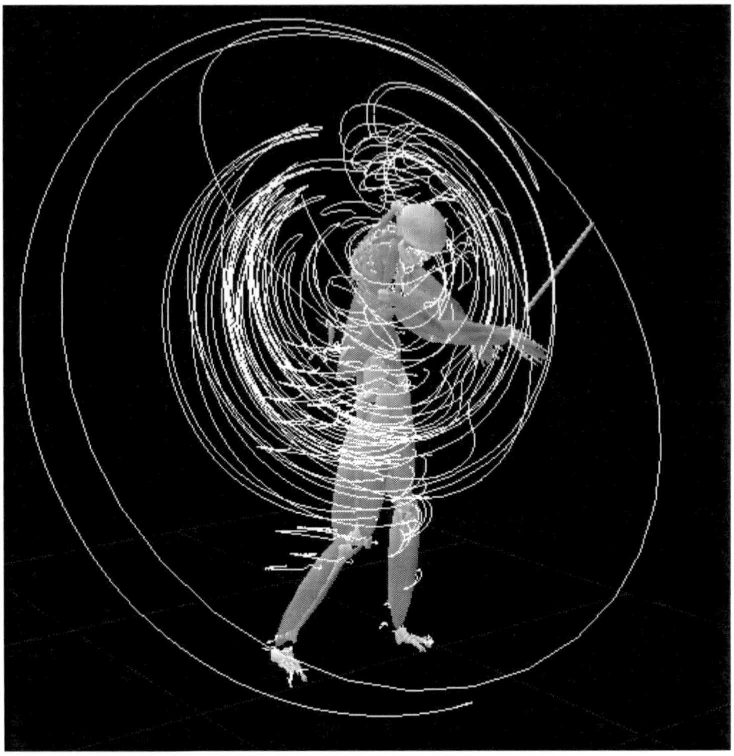

Figure 14. Signature of golf swing generated by using 3D motion capture technology.

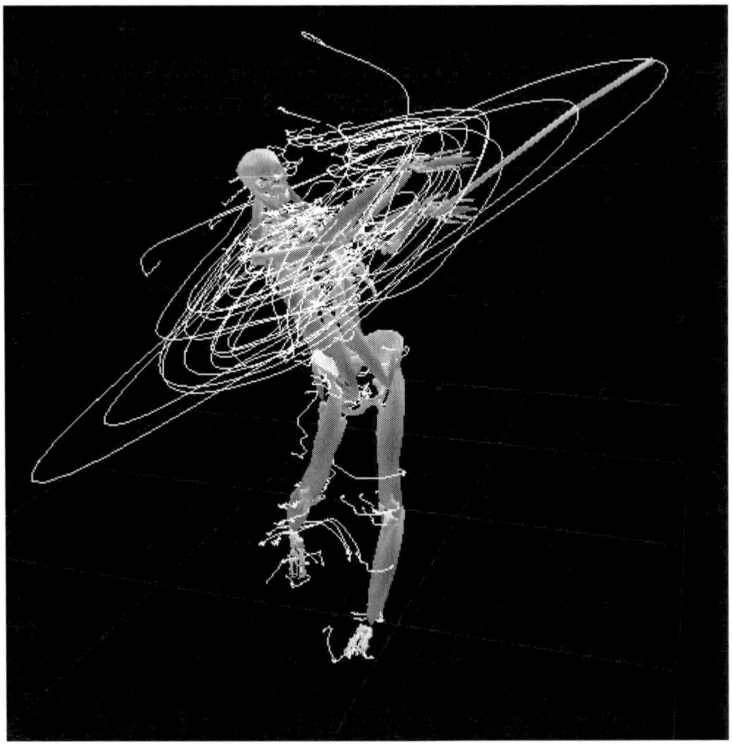

Figure 15. Signature of baseball swing.

## 3.2. Exemplary Movement Signatures Created by 3D Motion Capture Technology

Images generated by motion capture trajectories suggest factors allude to human experience. This characteristic is a key to establishing movement identity. (Shan et al., 2010). The language typically used to describe them is firmly rooted in metaphors of perceptual experience; the "signature" of each activity can be identified by weighting factors such as stability (balance, body position/posture), athleticism (weight, speed, power/vigor), quality of movement (control, smoothness/abruptness), and dimension of the movement (excursion). Selected images, discussed below, show how the use of motion capture technology can enhance discourse about perceptual characteristics of human movement.

*Example 1: Signature of Bicycle Kick in Soccer*

When performing a bicycle kick in soccer, a player jumps, is in airborne flight for the kick, and then falls to the ground. As the player's trunk reaches approximately the horizontal position, like pedaling the wheels of a bicycle, a player kicks the ball from above his/her head while falling. Perhaps the most memorable bicycle kick of all time was performed by Pelé from Brazil (IFFHS, 1999), the most famous soccer player of the 20th century (Figure 16).

The signature of a bicycle kick (Figure 17) created by using 3D motion capture technology shows the following to be main characteristics: trunk trajectories have relatively small excursions and they tended to be clumped in one area of the image. Largest trajectories (arms, legs and feet) visibly occur on a plane rather than in a volume. There is the clear presence of a center, circled by trunk trajectories. A centripetal center suggests weight and dynamic balancing (Figure 17).

*Athleticism*: Long trajectory lines of the legs, particularly the feet, contrast the "low-energy" shorter lines of the trunk, lending the impression of a speed differential and a directed power. Orientation of the foot and leg trajectories is in contrast to those of the rest of the image, showing dynamism.

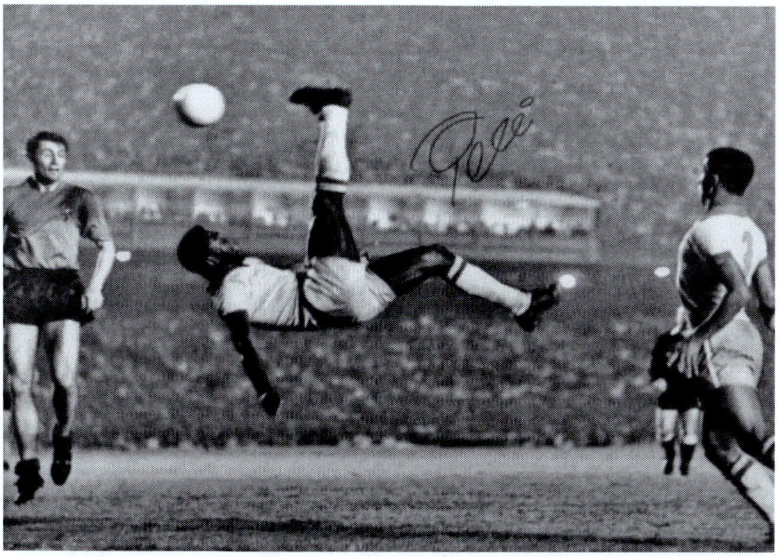

Figure 16. Bicycle kick performed by the soccer "artist" Pele in 1965 (Soccerzlife, 2008).

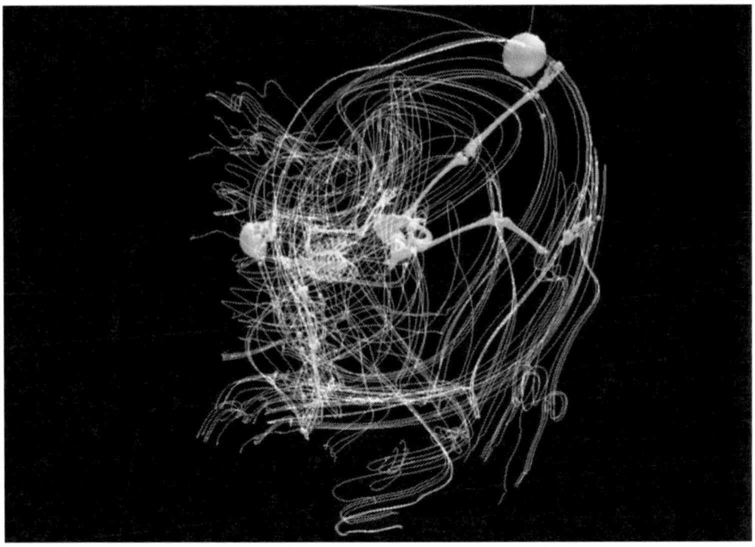

Figure 17. Signature of bicycle kick in soccer.

*Quality of movement*: Trajectories were continuous and smooth, showing controlled gestures. The acute angle change in direction of the ball indicates thresholds of the activity where energy of the movement was released.

*Dimension*: Generally compact excursions of trajectory lines shows the movement to be occurring in a limited space, suggesting bodily control.

*Example 2: Signature of a Hand Spring in Gymnastics*

Its key characteristics are as follow:

*Stability and quality*: A dense mass of trajectories, and a planar orientation in the central trajectories of the image are marks of balance and control (Figure 18). These contrast those distal trajectories that have a regular and curvilinear path, indicating speed and ease.

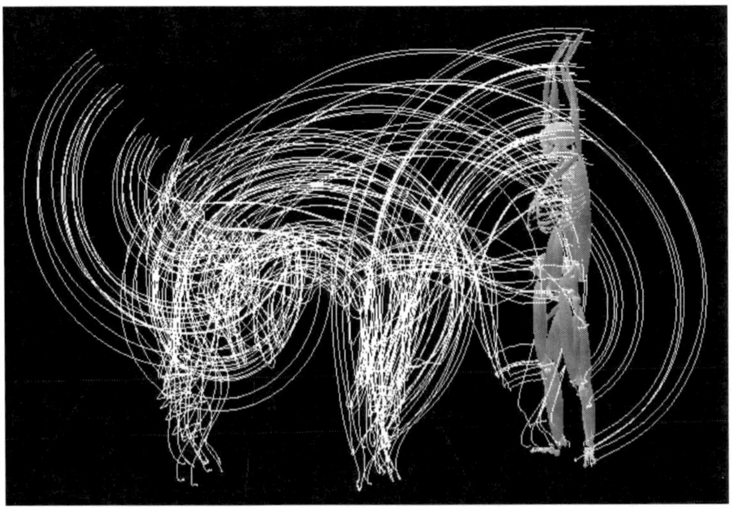

Figure 18. The signature of hand spring in gymnastics.

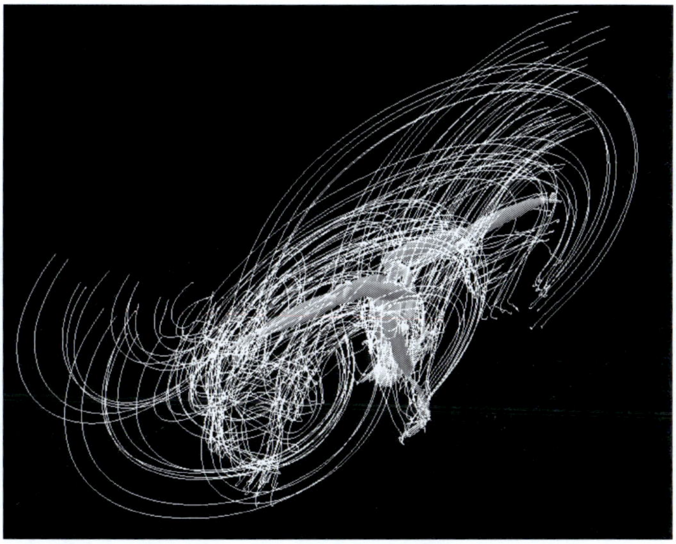

Figure 19. The signature of hand spring in gymnastics – a top view.

In contrast with the "explosive" bicycle kick, the smooth trajectories of the handspring indicate that the building and release of the body's kinetic energy are gradually exercised during the whole performance of the motor skill.

*Athleticism dynamism and dimension*: a top view of the activity emphasizes athleticism and dimension (Figure 19).

*Example 3: Signatures of Basic Ballet Skills*

From a biomechanics point of view, the movement characteristics of some ballet and gymnastics skills are similar to each other. The signatures of three selected ballet skills – fan kick (Figure 20), layout (Figure 21) and *Pas* (Figure 22) – are shown here. Like gymnastic skills, trajectories of ballet skills have a regularity of curvilinear paths, clearly indicating smooth, controlled gestures.

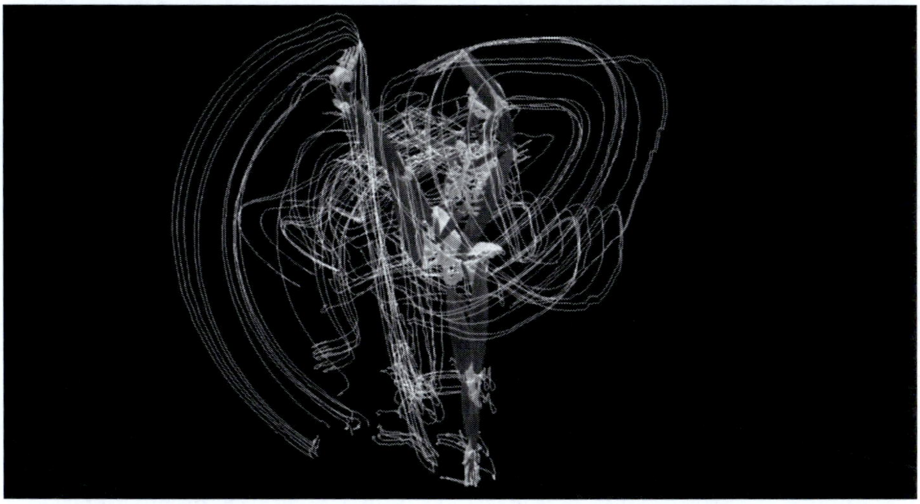

Figure 20. Signature of a ballet skill – Fan kick. (asymmetric is the common characteristic).

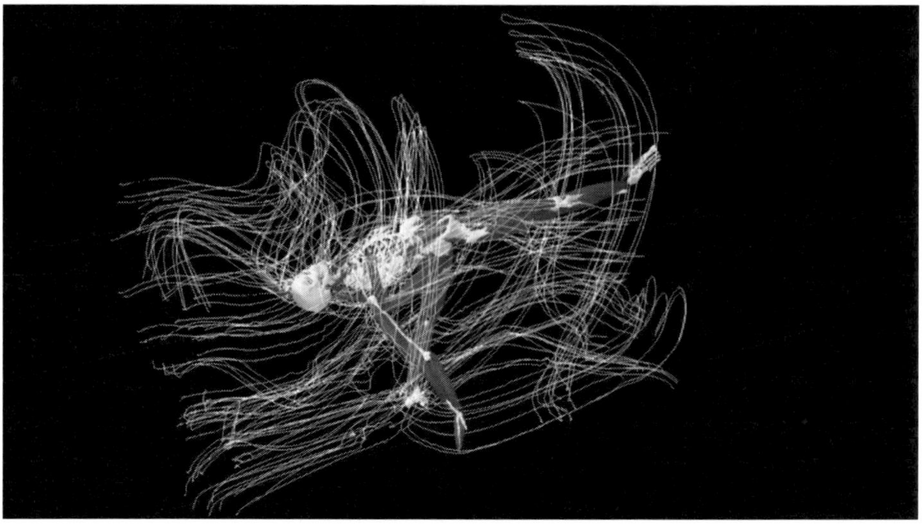

Figure 21. Signature of a ballet skill – layout.

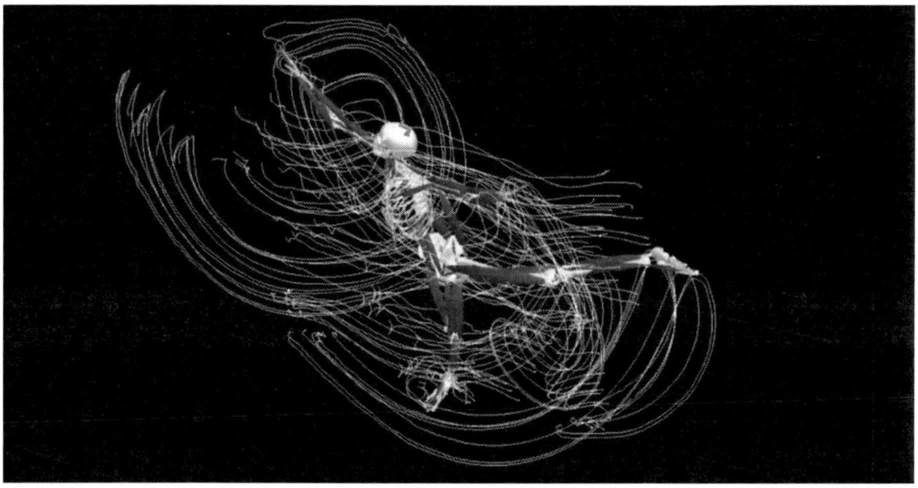

Figure 22. Signature of a ballet skill – pas.

Although different skills create different signatures, one common point of the signatures – asymmetry – is noticeably clear, signifying transfer in weight and dynamic stability during performance of skills.

*Example 4: Signatures of Fundamental Skills in Classical Indian Dance (Figure 23 – 26)*

In this art form, hand gestures and steps were executed according to highly codified traditions. In Indian dance, different combinations of gestures and movements provide a huge vocabulary through which stories can be depicted in great detail.

Compared to western dance (e.g., ballet), the Indian dance skills visibly show symmetry. The architectonic structure of trajectory lines show elements of symmetry, balance, and regularity of movement. Generally compact excursions of trajectory lines suggest movements in a limited space, i.e., Indian dance skills are less translational/dynamic that classical ballet skills.

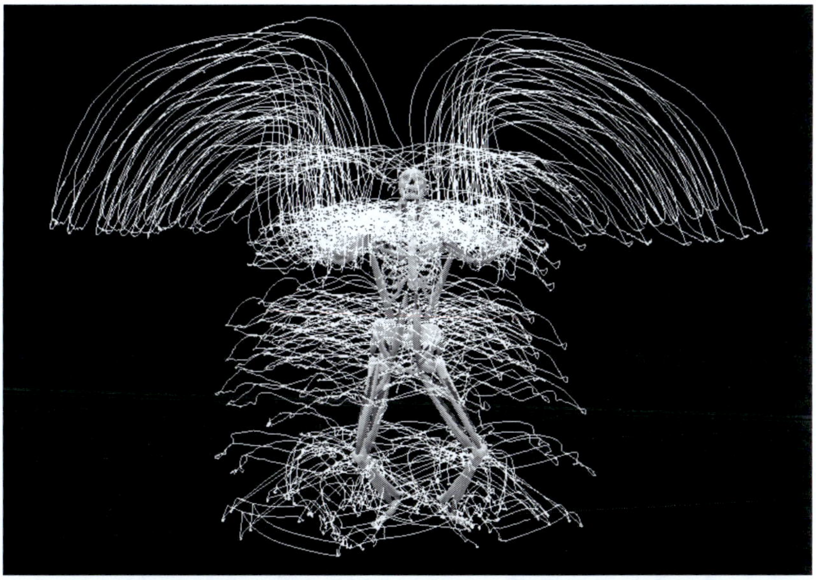

Figure 23. Signature of an India dance skill.

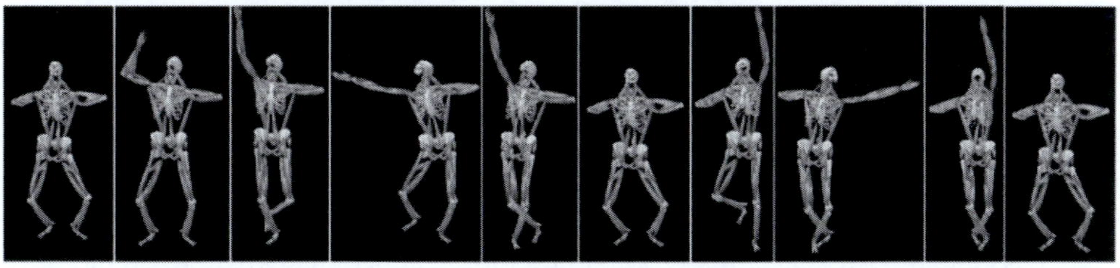

Figure 24. Sequential stop-frame reconstruction of the Indian dance skill in Figure 23.

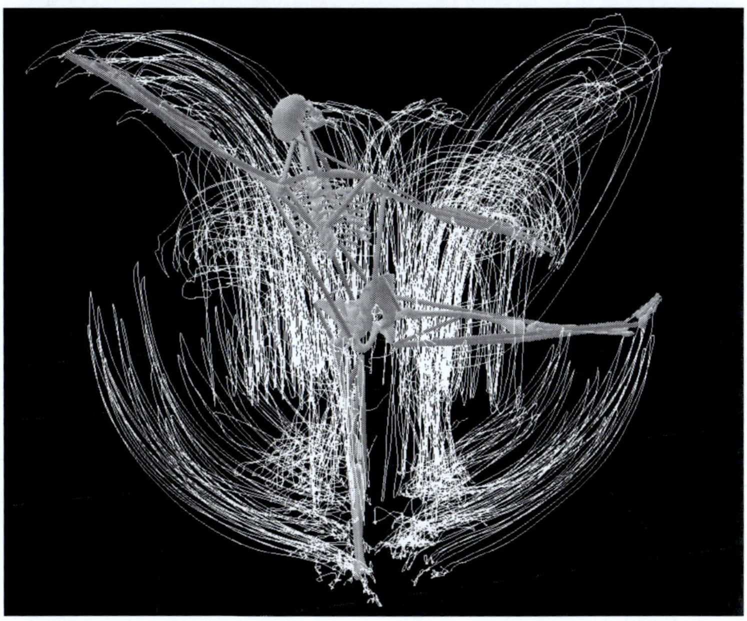

Figure 25. Signature of another India dance skill.

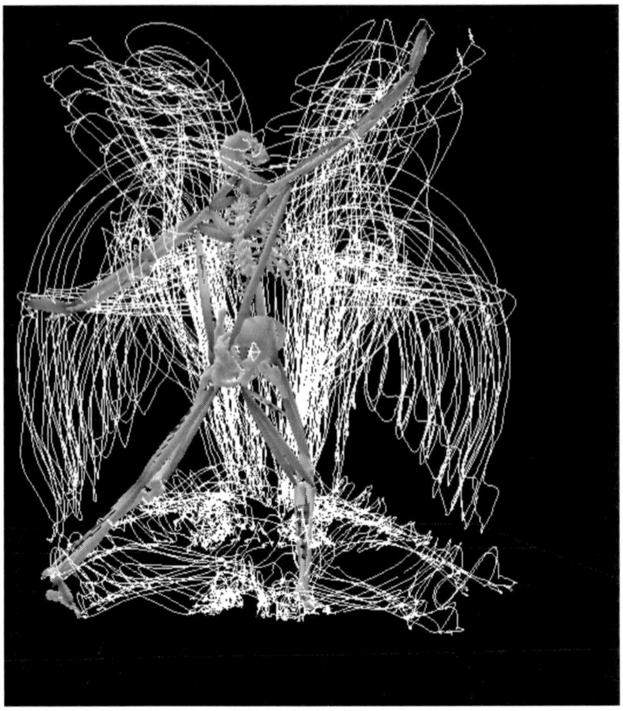

Figure 26. Signature of a 3rd Indian dance skill, selected for the study.

*Example 5: Signature of Boxing*

Depending on their goals, human motor skills entrainment can be divided into two general categories: movements that focus on regularity & consistency (e.g., basic motor skills in gymnastics and dancing) and movements that are irregular and seemingly improvisatory. The latter one is in response to events that are difficult to predict. A good example can be taken from boxing – a boxer has to keep his upper body continuously moving in an irregular way in order to avoid being punched and, at the same time, he needs to be able to deliver a punch whenever opportunity arises. Figure 27 and 28 show the signature of such a movement. The main characteristics of the skill are as follows:

*Stability*: Upper body trajectories had irregular, non-linear excursions, indicating movement that is trained to be unpredictable. The smallest trajectories (legs and feet) suggest that stability and balance during boxing are achieved primarily through control of the lower limbs.

*Athleticism*: The near-linear excursions of left arm and hand contrast the irregular lines of the upper body, lending the impression of a speed differential and a directed power. Orientation of the left arm and hand trajectories are in contrast to those of the rest of the image, showing the dynamism of the skill - explosive, fast and reactionary.

*Quality of movement*: The contrast between the irregular and regular trajectories gives some idea of the challenges of gestures control faced by a boxer. Abrupt changes in direction of left hand trajectories indicates thresholds of the activity or bodily limits where energy of the movement is released.

*Dimension*: Generally compact excursions of trajectory lines shows the movement to be occurring in a limited space indicating that, for all its unpredictability, it is highly controlled.

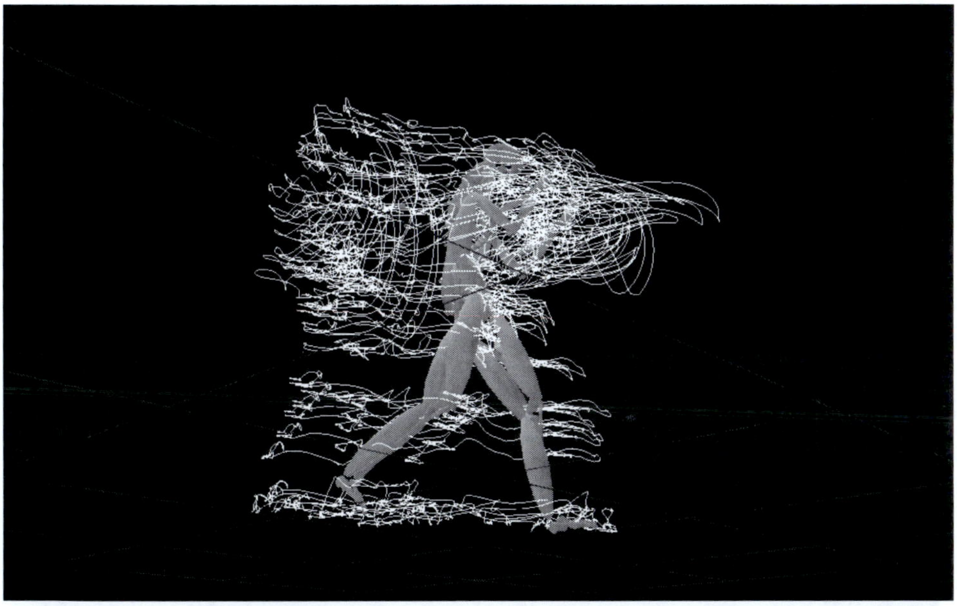

Figure 27. Signature of boxing punch (bottom view).

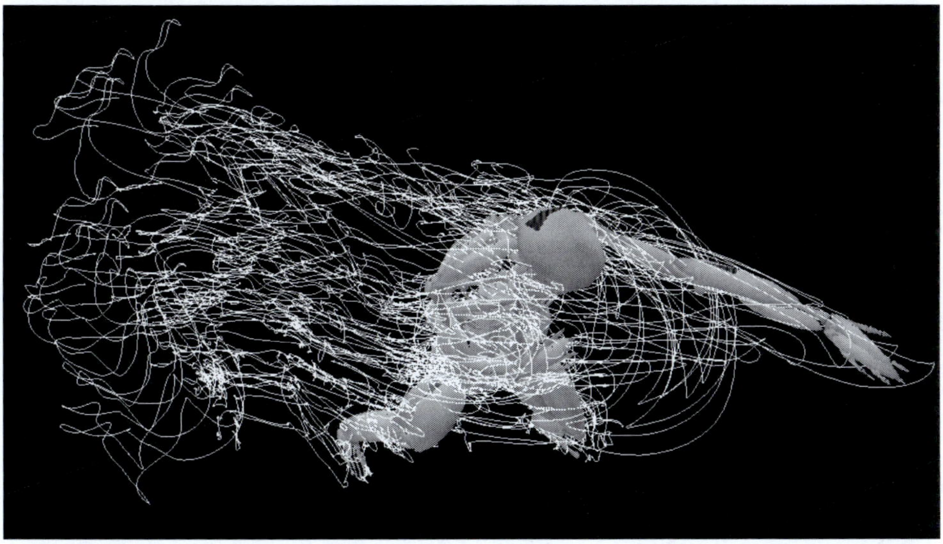

Figure 28. Signature of boxing punch (top view).

*Example 6: Signature of Ice Hockey Shooting (Figure 29 and 30)*

Humans interact with tools in many situations, including daily work, hobbies, and leisure activities. The tools in sports activities can be divided into three broad categories: 1) equipment for manipulating human motor effects (e.g., speed, force etc.), 2) equipment for preventing injury, and 3) equipment for creating performance effects. For the first category, sport equipment usually works as a lever, changing the speed advantage of the human-tool system, i.e., amplifying the movement achieved at the distal end of the tool compared to the amount of movement at the proximal end. Simply said, one can use leverage effects to propel an object faster (distal end) by applying a force at a slower speed (proximal end). Ice hockey, golf, baseball and lacrosse are some examples where sticks of one type or another are used for this purpose.

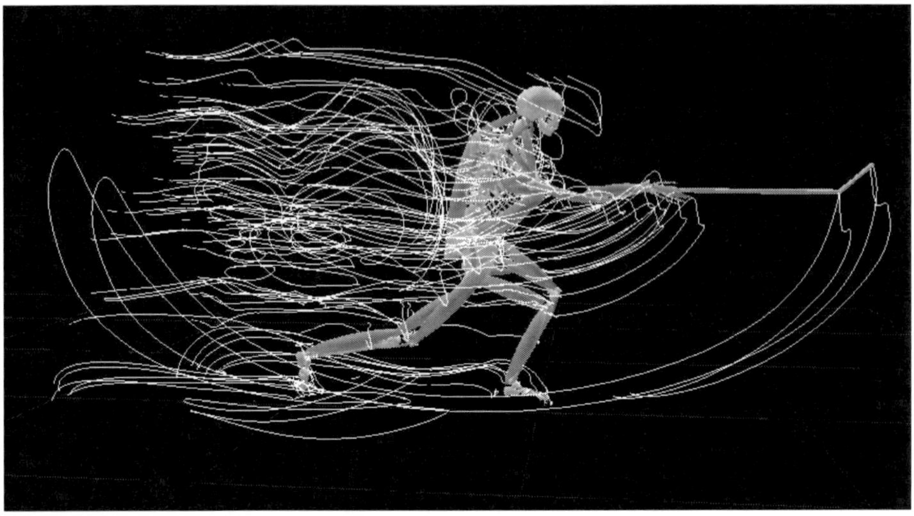

Figure 29. The signature of ice hockey shooting.

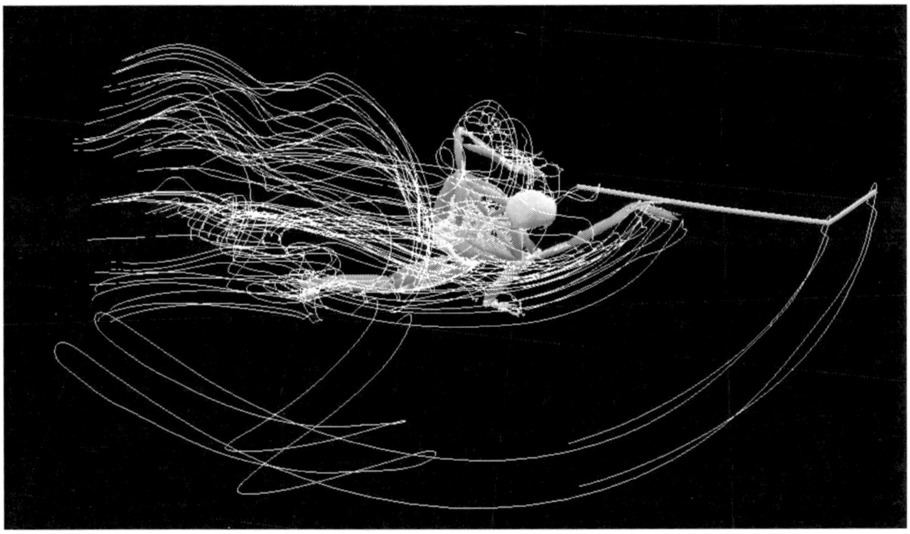

Figure 30. The signature of ice hockey shooting (top view).

The main features of the signature of ice hockey shooting are:

*Stability*: Most trajectories appear to be aligned in one general direction and they have nearly similar lengths. Relatively parallel to one another, trajectories (particularly trunk and head) show a state of controlled balance.

*Athleticism and dimension*: Swirling trajectories of the arms and the equipment are related to energy and athleticism. The double structure of these swirls indicates building and release phases of the movement. The contraction of the curve lengths between arms and stick indicate a notable difference in range of motion between the player's hands and the stick's blade. It also gives scale to the overall gesture – a larger movement dimension as a whole (Figure 30).

*Quality of movement*: Abrupt changes in direction of trajectories – points of curvilinear discontinuity – indicate the release of pent-up power, showing dynamism.

*Example 7: Signature of an Overhand Throw in Lacrosse*

Like ice hockey, the objective of the lacrosse is to score goals by shooting the ball into an opponent's goal. In lacrosse, a stick with a webbed head is used to catch, carry, pass and shoot the ball. The overhand throw is a fundamental skill (Figure 31).

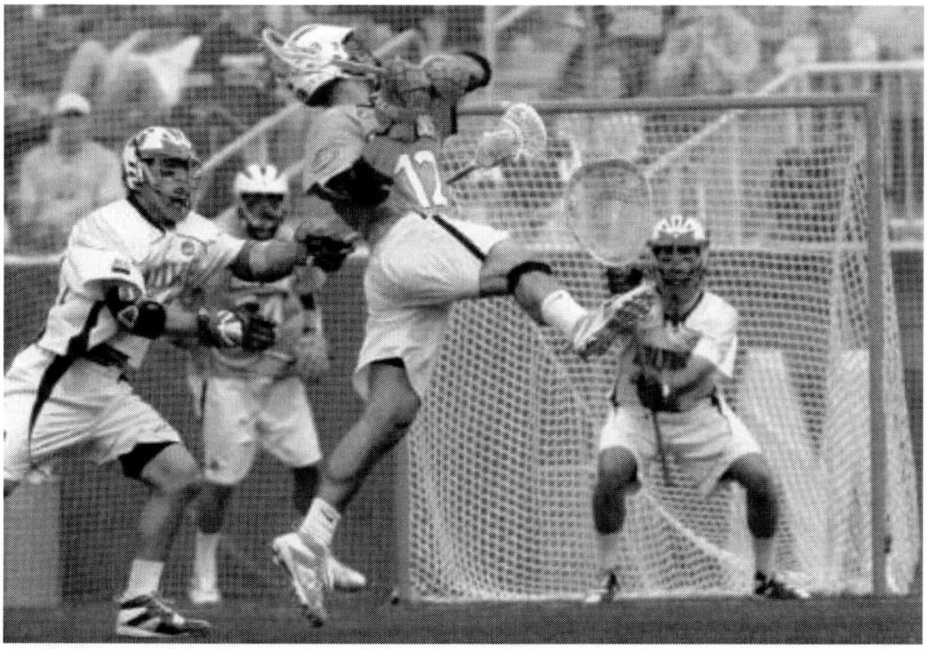

Figure 31. Lacrosse overhand throw (Rogash, 2008).

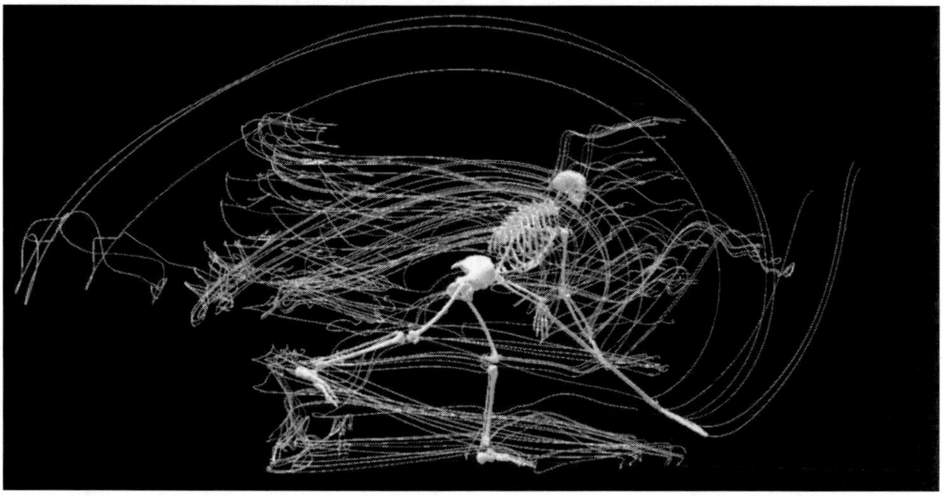

Figure 32. Signature of lacrosse overhand throw.

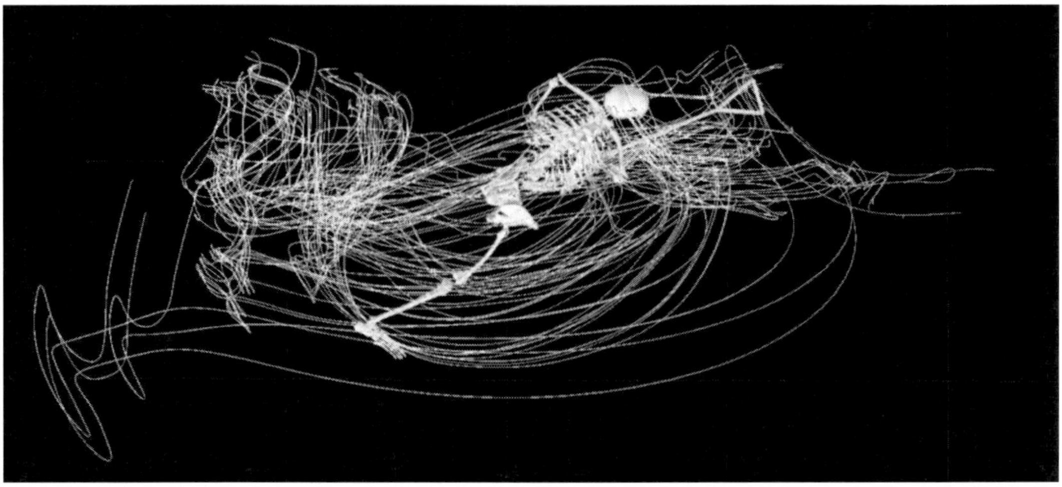

Figure 33. Signature of lacrosse overhand throw (top view).

Due to similarity in their basic tenets, it seems intuitive to expect the biomechanics of movement control in ice hockey and lacrosse to be analogous. However, a comparison of the two (Figures 29 & 30 and Figures 32 & 33), shows clear differences in: 1) movement dimension, 2) equipment orientation, and 3) speed.

These differences give the impression that the two games are played at different levels of speed. Again, movement signatures can not only inform us both space and temporal dimensions of human motor skills (an idea perhaps hinted in Da Vinci's drawing: the Vitruvian Man), but they also show differences of these two movement qualities among various activities. Such qualities can only be achieved through analysis of three dimensional data.

*Example 8: Signature of Throwing in American Football (Figure 34)*

Throwing is one of the basic motor functions that distinguish humans from most animals. Most species lack the basic anatomy needed to throw a ball, spear, or other projectile. American football is one activity that has required the development of specialized throwing techniques. The main characteristics of the movement signature are:

*Stability*: trunk trajectories have relatively small, curvilinear path and they tend to be clustered in one area of the image. Clear presence of a center, circled by rotational trunk trajectories suggests weight and balance. Trajectories of the legs and feet visibly occur on a plane rather than in a volume, suggesting a dynamic translation.

*Athleticism*: Relative linear trajectory lines of the feet and circular gestures of the hands suggest a building of linear and angular momentum during the throw, giving the impression of a speedy powerful movement. Orientation of leg, trunk and arm trajectories show dynamism.

*Quality of movement and dimension*: Continuity of trajectories indicates smooth, controlled gestures. Generally compact excursions of trajectory lines show small dimension of movement.

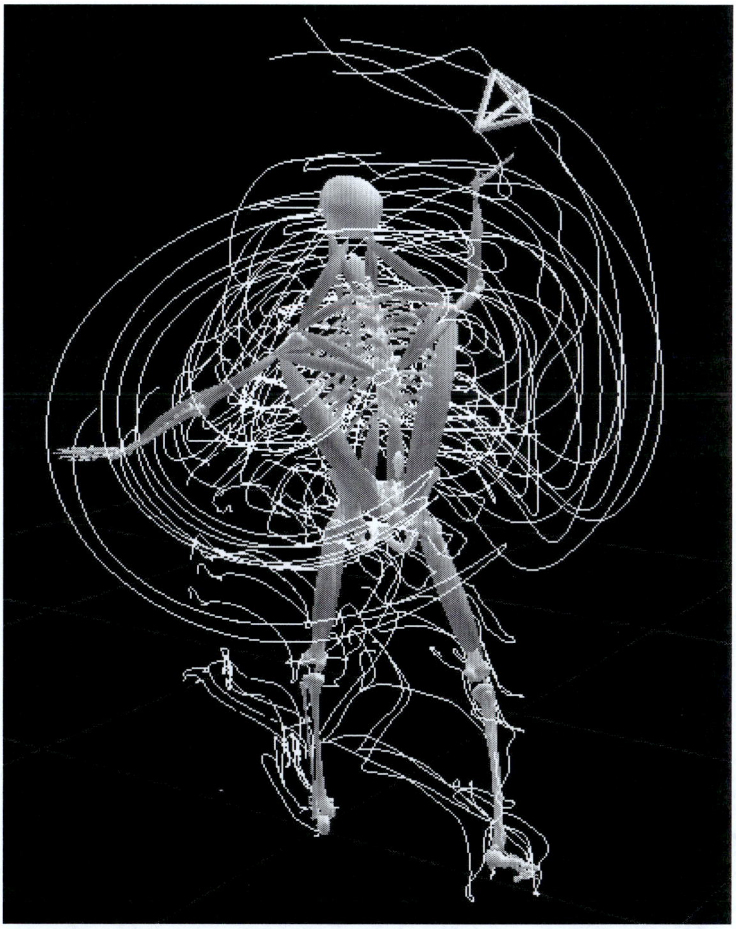

Figure 34. The signature of American football throwing.

*Example 9: Signature of Shot Put*

Example 8 gives the general characteristics of human throwing: translation of lower limbs for accelerating the throwing object and rotation of upper body and limbs for amplifying the effect of translation to maximize the release speed of the throwing object. However, it is often in the reality that an object is too heavy for the athlete to rotate it around his/her trunk as a means of increasing its speed. In such a situation, one has to use different motor control to maximizing release speed.

Shot put is one such sport. Figure 35 shows the characteristics of the movement: 1) facing backwards and using legs gliding (linear acceleration), 2) rotating 180° (asymmetrical control), left arm extended for maximizing the effect of the rotation, i.e., a second stage of accelerating the shot, then 3) tossing the shot (asymmetrical control), the right arm extension for finally accelerating the shot.

The signature of shot put (Figure 36) clearly shows the characteristics described above: 1) the leg and foot trajectories visibly occur on a plane rather than in a volume, suggesting positional translation; 2) a contraction of the curve lengths between left and right arm indicate a notable difference in range of motion between the arms, showing an asymmetrical control of the arms; 3) relative linear trajectory lines of right hand show a creation of linear momentum during the whole movement, lending the impression of a speedy powerful movement. The contrast between right and left sides of the body trajectories reveals a typical asymmetrical motor control skill.

2-D images that trace movement trajectories communicate something of bodiliness to the extent that, as abstractions, they make tangible qualities that appeal to a viewer's familiarity with the human condition. The experience of the viewer bridges any gaps between perception and understanding of images representative of human movement. This experience imbues trajectory data with meaning, providing a third dimension to a still image, even without a referential inclusion of a human form. The following three examples may be used to demonstrate this point.

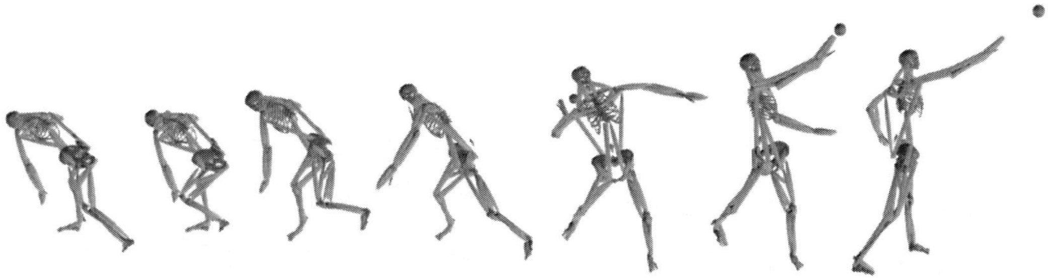

Figure 35. Sequential stop-frame reconstruction of the movement.

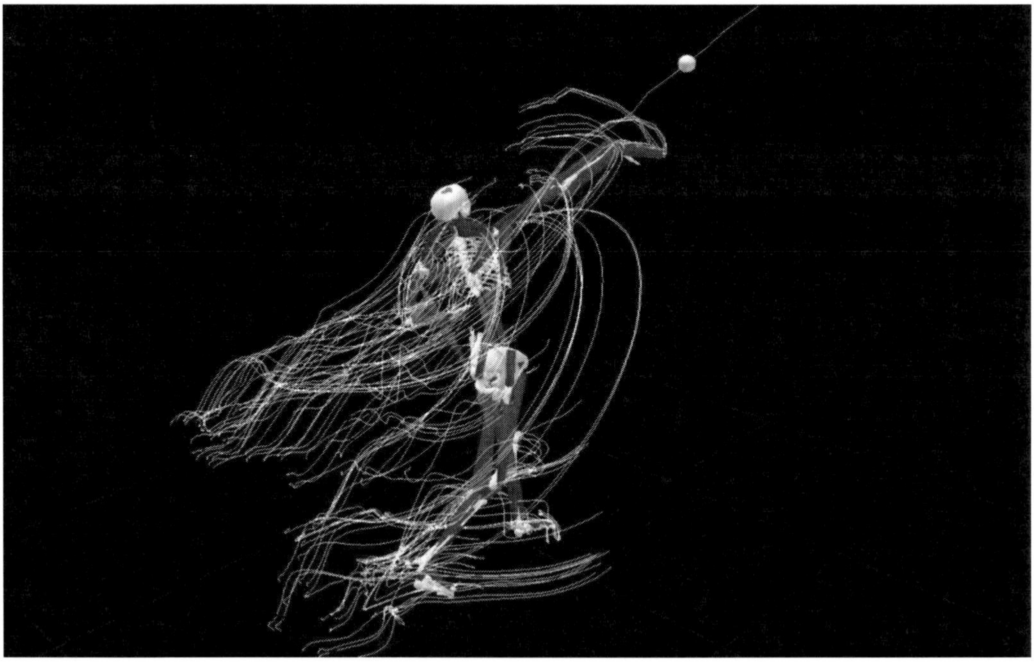

Figure 36. The signature of shot put.

*Example 10: Signature of Lifting an Object*

Lifting is a basic human motor skills that nearly everyone has experienced. Therefore, even with the movement signature stripped of its human-form reference figure, the essential characteristics of the signature are clear (Figure 37, right).

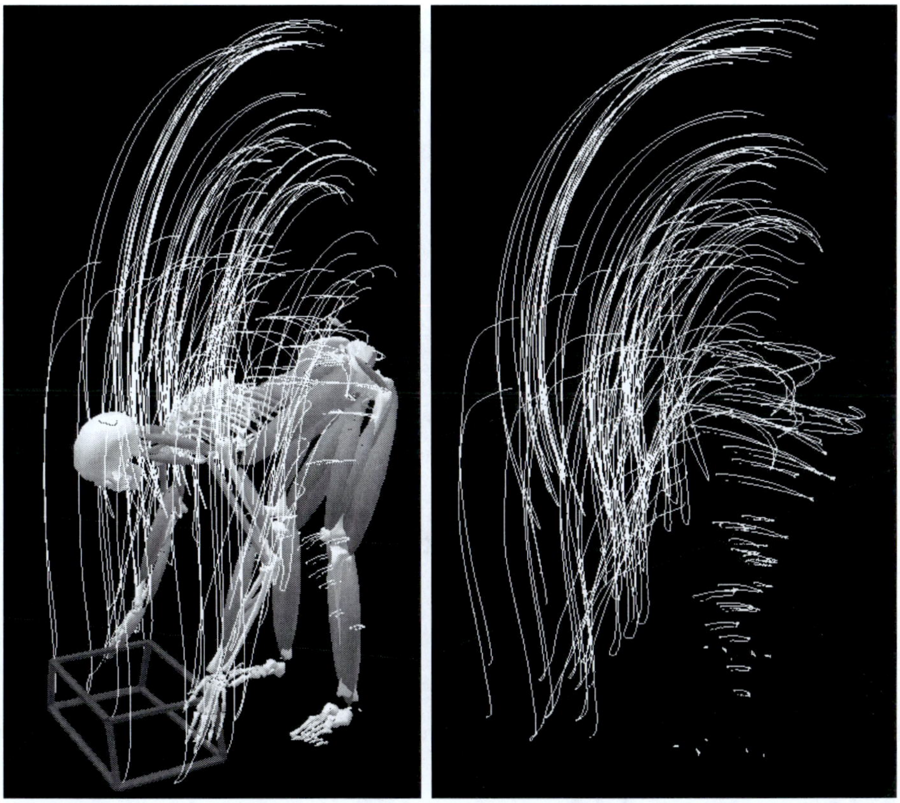

Figure 37. Signature of lifting in daily life (left) and trajectory images represent the activity (right).

*Example 11: Signature of 3-Flip in Skateboarding*

Skateboarding is an action sport which involves riding and performing tricks using a skateboard. A 3-flip is a type of skateboarding trick in which the skateboard rotates around its horizontal and vertical axis simultaneously. This motor skill involves: 1) the board moving with skateboarder as he jumps, 2) the board rotating under the rider's feet during the airborne phase, and 3) toward the end of this rotation, the skater catches the board with his/her feet and lands on the ground. A time-based execution of the skill created by using 3D motion capture technology is shown in Figure 38. The signature of the skill is presented in Figure 39. The signature visibly shows the jump of the skateboarder, separation from and catching of the board, and the simultaneous rotation of the board around horizontal and vertical axes during the airborne phase. When the referential human form is removed from the image, the essential characteristics of the movement signature are still clear (Figure 40).

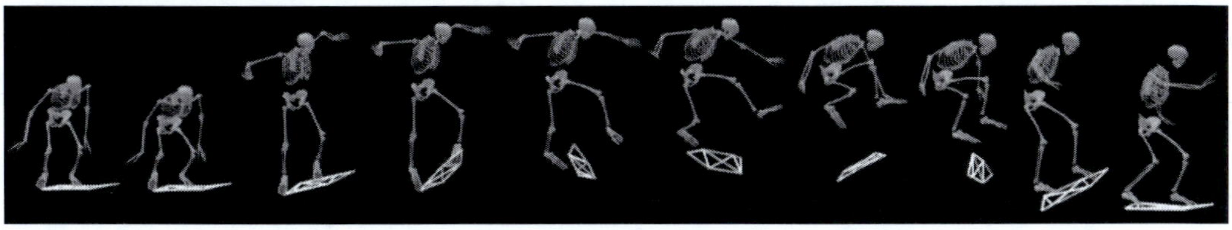

Figure 38. Sequential stop-frame reconstruction of the movement.

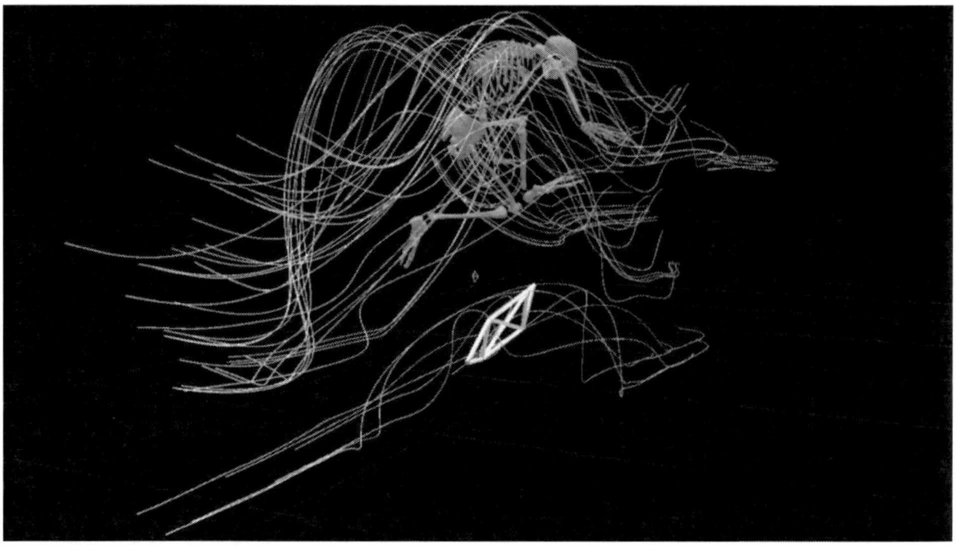

Figure 39. Signature of 3 flip in skateboarding.

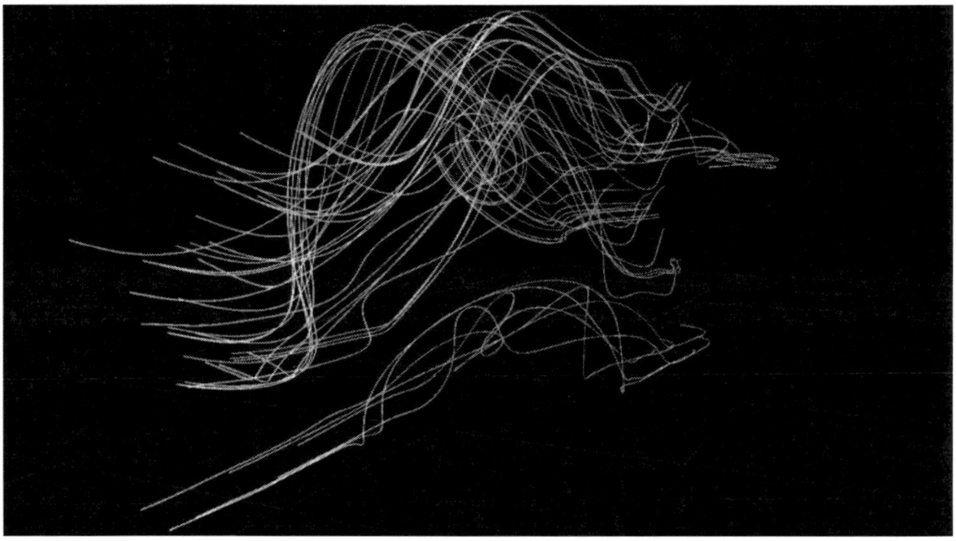

Figure 40. Trajectory images represent the activity of 3 flip in skateboarding.

*Example 12: Signature of Conducting*

Music conducting involves the communication of temporal ideas using non-verbal cues. Such cues are typically exposed by a conductor's body language as well as the tool he/she uses – a baton. Therefore, the trajectories related to baton movement form one core element of the communication process. Figure 41 shows a brief excerpt from the activity of conducting. The clearly repetitive structure of trajectory lines showed controlled, directed movement. Limited range of movement in the body and right arm (center & right of the image) expressed a nearly static posture, while curvilinear trajectories at the proximal and distal ends of baton-arm chain (left) showed an element of leverage effects, i.e., increasing range of motion of a baton tip in order to maximize a conductor's gestural cues.

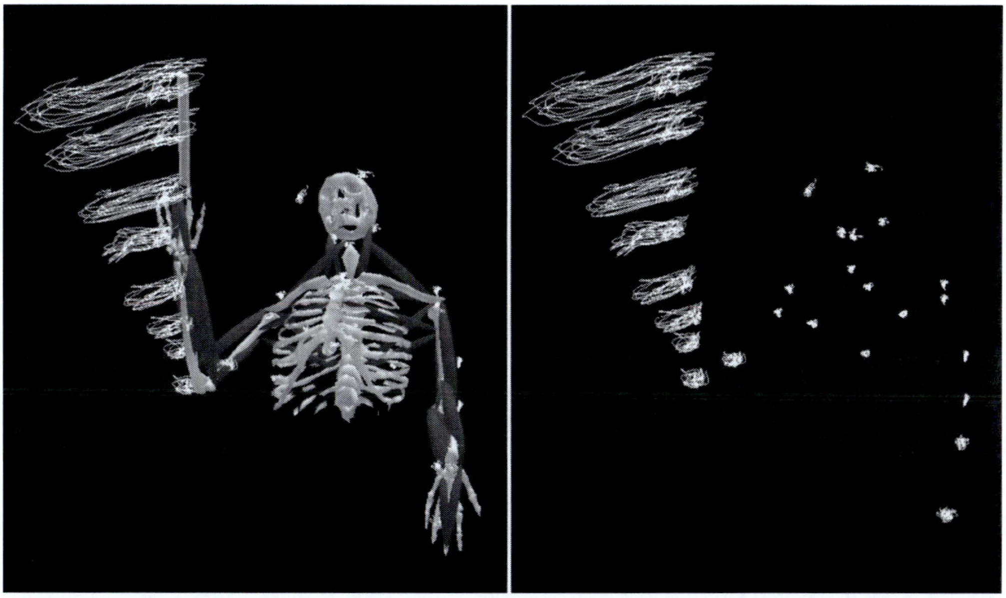

Figure 41. Signature of conducting (left) and trajectory images represent the activity (right).

There is a clear asymmetry in the image trajectories, a quality that reinforced the impression of a bodily function being connected to the use of a tool in the service of larger mechanism. Continuous gesture trajectories with no obvious points of curvilinear irregularity indicated dynamism to be relatively unimportant. Energy expenditure of right-arm is directed toward control. Generally compact excursions of trajectory lines suggested movements in a limited space or a limited time.

Using 3D motion capture technology, the current paper provides exemplary movement signatures from selected activities. Uniqueness, that which can be used to identify signature characteristics, can be defined by qualities such as stability, balance, athleticism, power, symmetry and movement control. As such, linear trajectories can aid discourse about perceptual characteristics of human movement, its functionality and creativity, and its ability to communicate through metaphor and commonality of experience. In the case of imagery that captures perceptual characteristics of human movement, objectification of the nature of movement signatures provides one more way to comprehend human gesture as both aesthetic and as communicating embedded experiential and cultural dynamics.

## CONCLUSION

Historical documentation can be used to suggest that the study of biomechanics begins with artworks documenting components of the human musculoskeletal system – the artworks of biomechanics. Da Vinci considered and explored interactions between science and art as inseparable ideologies. “No science (in the Renaissance) was left untouched by the new modes of illustration, whether the descriptive science of anatomy or the three-dimensional geometrics of cosmology” (Kemp, 2005). With regard to human movement science, da Vinci’s work has been particularly important. He created artworks that vividly show scientific contents related to the human motor system. About a hundred years later, Borelli’s work further demonstrated the power of art as a means for disseminating quantitative scientific knowledge. However, due to limitations of technology, affective artistic representations of human movement prior to the 19th century remained mostly mimetic, poetic depictions of gesture on a composed 2D picture plane. Affects representative of movement relied on the viewer to understand iconography, metaphor or other cultural loading present in the work. Such artworks might be categorized as movement sketches, because information from the third dimension of Cartesian space and the temporal dimension of human experience are not fully articulated. What these visual artists did have at their disposal was a set of skills and traditions directed by a disciplined eye. Muybridge's multi-exposure photographs successfully included both space and temporal characteristics into single-frame images. Many of his artworks can be understood as identifying elements of movement signatures. Muybridge's photograph techniques limited him strictly to 2D exploration. In the works of artists such as Degas and Duchamps, we can see the artist's eye informing and manipulating photographic principles to include time and additional dimension.

In the current paper, 3D motion capture is used to generate signatures of human movement that are objective and representative of selected activities. 3D motion capture is a technology specifically designed for time-based documentation of movement. The current article shows that the technology can effectively synthesize time-based trajectory information into single-frame, time-independent images. Such visual images imply space and temporal characteristics as well as unique combinations of elements that can be thought of as signatures of human movement. Their ability to objectively represent key characteristics informs viewers' perceptions and stimulates interpretation. The discussion above argues that these images contain latent and communicative content - visual cues related to perception, cognition, and cultural practice that offer an effective metaphor for human experience.

## References


Abbott, A., & Rutherford, A. (2005). Artists on science: scientists on art. *Nature, 434*(7031), 293.

Arnold, V. I. (1989). *Mathematical Methods of Classical Mechanics* (Second ed.). New York: Springer.

Bohr, N. (1928). The Quantum Postulate and the Recent Development of Atomic Theory. *Nature (Supplement), 121*, 580-590.

Ethier, C. R., & Simmons, C. A. (2007). *Introductory Biomechanics - From Cells to Organisms*. Cambridge: Cambridge University Press.

Gonpertz, W. (2012). *What are you looking at?* New York: Dutton, Penguin group.

Hatze, H. (1974). The Meaning of the Term "Biomechanics". *Journal of Biomechanics, 7*(2), 189-190.

Hay, J. G. (1985). *The Biomechanics of Sports Techniques* (Third Edition ed.). Englewood Cliffs: Pretice-Hall Inc.

Hunter, S., Jacobus, J., & Wheeler, D. (2004). *Modern Art* (Third Edition Revised and Expanded ed.). New York: Prentice-Hall.

IFFHS. (1999). *The World's best Player of the Century* Retrieved September 16, 2013, from http://www.iffhs.de/?42d03e32a16f43809fa3c17c13c09e20a15ff3c09f32b17f7370eff3702bb1c2bbb6e28fc380de43110f83d00e09

Kemp, M. (2005). From science in art to the art of science. *Nature, 434*(7031), 308-309.

Martin, R. B. (1999). *A Genealogy of Biomechanics*. Paper presented at the 23rd Annual Conference of the American Society of Biomechanics, Pittsburgh.

Pope, M. H. (2005). Giovanni Alfonso Borelli--the father of biomechanics. *Spine, 30*(20), 2350-1355.

Rogash, J. (2008). Johns Hopkins photos Retrieved October 18, 2013, from http://i.usatoday.net/sports/gallery/2008/05/24/s080524_lacrosse.jpg

Shan, G., Visentin, P., & Harnett, T. (2010). A Novel Use of 3D Motion Capture: Creating Conceptual Links between Technology and Representation of Human Gesture in the Visual Arts. *Leonardo, 43*(1), 34-42. doi: 10.1162/leon.2010.43.1.34

Shan, G., Zhang, X., Li, X., Hao, W., & Witte, K. (2011). Quantification of Golfer-club Interaction and Club-type's Affect on Dynamic Balance during a Golf Swing. *International Journal of Performance Analysis in Sport, 11*(3), 417-426.

Siegel, E. (2012). *The Physics of Happy Gilmore*, Retrieved September 16, 2013, from http://scienceblogs.com/startswithabang/2012/09/15/weekend-diversion-the-physics-of-happy-gilmore/

Soccerzlife. (2008). Overhead goal by Coruch this weekend. *Get Your Goals Right* Retrieved September 19, 2013, from http://soccerzlife.wordpress.com/2008/10/07/overhead-goal-by-coruch-this-weekend/

Solnit, R. (2003). *River of Shadows: Eadward Muybridge and the Technological Wild West*. New York: Viking Penguin.

Veltman, K. (1992). Leonardo da Vinci: Studies of the Human Body and Principles of Anatomy. In K. Schreiner & N. Schnitzler (Eds.), *Gepeinigt, begehrt, vergessen. Symbolik und Sozialbezug des Körpers im späten Mittelalter und in der frühen Neuzeit* (pp. 287-308). Bad Homburg: Werner Reimers Stiftung.

Zhang, X. (2012). Muybridge, Motion Analysis and Art. *Arts BioMechanics, 1*(2), 105-114.

# INDEX

## D



## E



## F

## G



## H



## I

## J



## K



## L



## M



## N



## O



## P

## T



## U



## V



## W



## X



## Y

## Related Nova Publications

### HANDBOOK OF RESEARCH FOR BIOMECHANICS AND ANTHROPOMETRY IN SPORT SCIENCE

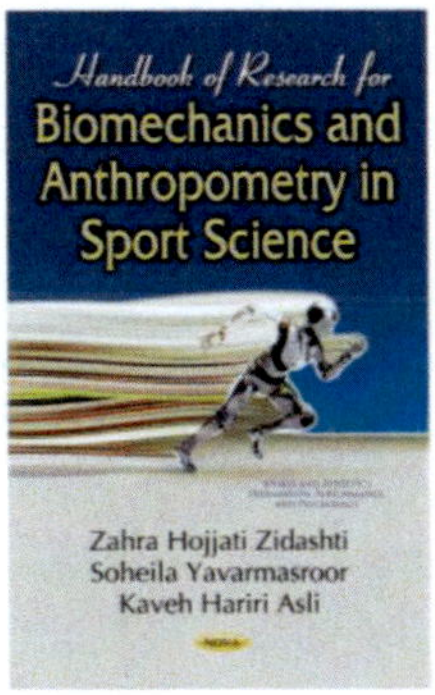


**AUTHORS:** Zahra Hojjati Zidashti, Ph.D., Soheila Yavarmasroor, and Kaveh Hariri Asli, Ph.D.

**SERIES:** Sports and Athletics Preparation, Performance, and Psychology

**BOOK DESCRIPTION:** This book provides new methods to novel applications of existing methods to gain understanding of the material and/or structural behavior of new and advanced systems. The authors provide innovative chapters on the growth of educational, scientific, and research activities among athletes and provides a medium for mutual communication between international sports academia.

**HARDCOVER ISBN:** 978-1-53610-466-0
**RETAIL PRICE:** $160

### APPLICATIONS IN ARTS BIOMECHANICS

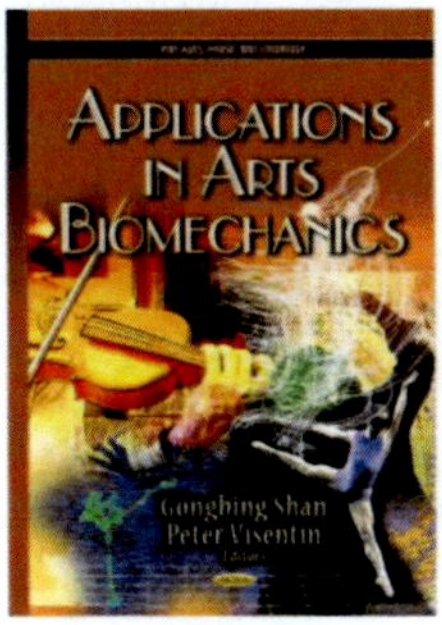


**EDITORS:** Gongbing Shan and Peter Visentin

**SERIES:** Fine Arts, Music and Literature

**BOOK DESCRIPTION:** "Applications in Arts Biomechanics" contains ten chapters covering music, dance, and the visual and media arts. It is organized in four parts: original research in each of the above plus a review article section.

**HARDCOVER ISBN:** 978-1-62808-015-5
**RETAIL PRICE:** $205

*To see a complete list of Nova publications, please visit our website at www.novapublishers.com*